Lecture Notes in Physics

Editorial Board

H. Araki, Kyoto, Japan
R. Beig, Vienna, Austria
J. Ehlers, Potsdam, Germany
U. Frisch, Nice, France
K. Hepp, Zürich, Switzerland
R. L. Jaffe, Cambridge, MA, USA
R. Kippenhahn, Göttingen, Germany
H. A. Weidenmüller, Heidelberg, Germany
J. Wess, München, Germany
J. Zittartz, Köln, Germany

Managing Editor

W. Beiglböck
Assisted by Mrs. Sabine Lehr
c/o Springer-Verlag, Physics Editorial Department II
Tiergartenstrasse 17, D-69121 Heidelberg, Germany

Springer
Berlin
Heidelberg
New York
Barcelona
Budapest
Hong Kong
London
Milan
Paris
Santa Clara
Singapore
Tokyo

The Editorial Policy for Proceedings

The series Lecture Notes in Physics reports new developments in physical research and teaching – quickly, informally, and at a high level. The proceedings to be considered for publication in this series should be limited to only a few areas of research, and these should be closely related to each other. The contributions should be of a high standard and should avoid lengthy redraftings of papers already published or about to be published elsewhere. As a whole, the proceedings should aim for a balanced presentation of the theme of the conference including a description of the techniques used and enough motivation for a broad readership. It should not be assumed that the published proceedings must reflect the conference in its entirety. (A listing or abstracts of papers presented at the meeting but not included in the proceedings could be added as an appendix.)
When applying for publication in the series Lecture Notes in Physics the volume's editor(s) should submit sufficient material to enable the series editors and their referees to make a fairly accurate evaluation (e.g. a complete list of speakers and titles of papers to be presented and abstracts). If, based on this information, the proceedings are (tentatively) accepted, the volume's editor(s), whose name(s) will appear on the title pages, should select the papers suitable for publication and have them refereed (as for a journal) when appropriate. As a rule discussions will not be accepted. The series editors and Springer-Verlag will normally not interfere with the detailed editing except in fairly obvious cases or on technical matters.
Final acceptance is expressed by the series editor in charge, in consultation with Springer-Verlag only after receiving the complete manuscript. It might help to send a copy of the authors' manuscripts in advance to the editor in charge to discuss possible revisions with him. As a general rule, the series editor will confirm his tentative acceptance if the final manuscript corresponds to the original concept discussed, if the quality of the contribution meets the requirements of the series, and if the final size of the manuscript does not greatly exceed the number of pages originally agreed upon. The manuscript should be forwarded to Springer-Verlag shortly after the meeting. In cases of extreme delay (more than six months after the conference) the series editors will check once more the timeliness of the papers. Therefore, the volume's editor(s) should establish strict deadlines, or collect the articles during the conference and have them revised on the spot. If a delay is unavoidable, one should encourage the authors to update their contributions if appropriate. The editors of proceedings are strongly advised to inform contributors about these points at an early stage.
The final manuscript should contain a table of contents and an informative introduction accessible also to readers not particularly familiar with the topic of the conference. The contributions should be in English. The volume's editor(s) should check the contributions for the correct use of language. At Springer-Verlag only the prefaces will be checked by a copy-editor for language and style. Grave linguistic or technical shortcomings may lead to the rejection of contributions by the series editors. A conference report should not exceed a total of 500 pages. Keeping the size within this bound should be achieved by a stricter selection of articles and not by imposing an upper limit to the length of the individual papers. Editors receive jointly 30 complimentary copies of their book. They are entitled to purchase further copies of their book at a reduced rate. As a rule no reprints of individual contributions can be supplied. No royalty is paid on Lecture Notes in Physics volumes. Commitment to publish is made by letter of interest rather than by signing a formal contract. Springer-Verlag secures the copyright for each volume.

The Production Process

The books are hardbound, and the publisher will select quality paper appropriate to the needs of the author(s). Publication time is about ten weeks. More than twenty years of experience guarantee authors the best possible service. To reach the goal of rapid publication at a low price the technique of photographic reproduction from a camera-ready manuscript was chosen. This process shifts the main responsibility for the technical quality considerably from the publisher to the authors. We therefore urge all authors and editors of proceedings to observe very carefully the essentials for the preparation of camera-ready manuscripts, which we will supply on request. This applies especially to the quality of figures and halftones submitted for publication. In addition, it might be useful to look at some of the volumes already published. As a special service, we offer free of charge LaTeX and TeX macro packages to format the text according to Springer-Verlag's quality requirements. We strongly recommend that you make use of this offer, since the result will be a book of considerably improved technical quality. To avoid mistakes and time-consuming correspondence during the production period the conference editors should request special instructions from the publisher well before the beginning of the conference. Manuscripts not meeting the technical standard of the series will have to be returned for improvement.

For further information please contact Springer-Verlag, Physics Editorial Department II, Tiergartenstrasse 17, D-69121 Heidelberg, Germany

János Kertész   Imre Kondor (Eds.)

Advances in Computer Simulation

Lectures Held at the Eötvös Summer School
in Budapest, Hungary, 16–20 July 1996

 Springer

Editors

János Kertész
Institute of Physics, Technical University Budapest
Budafoki út 8
H-1111 Budapest, Hungary

Imre Kondor
Bolyai College, Eötvös University
Amerikai út 96
H-1145 Budapest, Hungary

Cataloging-in-Publication Data applied for.

Die Deutsche Bibliothek - CIP-Einheitsaufnahme

Advances in computer simulation : lectures held at the Eötvös summer school at Budapest, Hungary, 16 - 20 July 1996 / János Kertész ; Imre Kondor (ed.). - Berlin ; Heidelberg ; New York ; Barcelona ; Budapest ; Hong Kong ; London ; Milan ; Paris ; Santa Clara ; Singapore ; Tokyo : Springer, 1996
 (Lecture notes in physics ; 501)
 ISBN 3-540-63942-X

ISSN 0075-8450
ISBN 3-540-63942-X Springer-Verlag Berlin Heidelberg New York

This work is subject to copyright. All rights are reserved, whether the whole or part of the material is concerned, specifically the rights of translation, reprinting, re-use of illustrations, recitation, broadcasting, reproduction on microfilms or in any other way, and storage in data banks. Duplication of this publication or parts thereof is permitted only under the provisions of the German Copyright Law of September 9, 1965, in its current version, and permission for use must always be obtained from Springer-Verlag. Violations are liable for prosecution under the German Copyright Law.

© Springer-Verlag Berlin Heidelberg 1998
Printed in Germany

The use of general descriptive names, registered names, trademarks, etc. in this publication does not imply, even in the absence of a specific statement, that such names are exempt from the relevant protective laws and regulations and therefore free for general use.

Typesetting: Camera-ready by the authors/editors
Cover design: *design & production* GmbH, Heidelberg
SPIN: 10644034   55/3144-543210 - Printed on acid-free paper

Preface

The dramatic development in computation has changed physics substantially. Besides the traditional methodological branches of the discipline, experimental and theoretical physics, a third branch has emerged: Computational Physics. This is not to say that computers are not used in other branches of physics as well, but in computational physics they have become the *main instrument* for exploring the laws of nature.

Statistical physics is one of the fields where the impact of modern computers has been enormous. Simulations of many-body systems has become a standard tool for research. In several classes of problems it is crucial to push the number of particles simulated to the limit, in particular, when there is no characteristic length in the system, such as for scaling phenomena, or when there are many characteristic lengths as, for example, in the case of granular systems. An important lesson learnt during the last two decades is that improving our simulation techniques via physical insight is at least as important as the experienced faster-than-exponential growth in hardware capacity for the purposes of achieving large-scale simulations.

Advances in computer simulation was a meeting held in Budapest in July 1996, as one of the series of Eötvös Summer Schools. After a general introduction, the discussion was centered around recent developments in physics-based algorithmic advances. The main emphasis of the school was on Monte Carlo techniques applied to many body systems, but topics from molecular dynamics, stochastic differential equations and optimization were also touched upon. Both lecturers and students found the school so useful that we decided to publish the lectures – resulting in the present volume. We hope it will contribute to the spreading of contemporary ideas in simulational statistical physics.

Finally, we would like to thank B. Bakó, I. M. Kondor, and T. N. Townsend for their help with the editing of this volume; and the Soros Foundation, the International Workshop on Theoretical Physics, and the Hungarian Academy of Sciences for their financial assistance towards the school.

Budapest, July 1997

János Kertész
Imre Kondor

Table of Contents

Introduction to Monte Carlo Algorithms
Werner Krauth 1

Cluster Algorithms
Ferenc Niedermayer 36

Optimized Monte Carlo Methods
Enzo Marinari 50

Monte Carlo on Parallel and Vector Computers
Dietrich Stauffer 82

Error Estimates on Averages of Correlated Data
Henrik Flyvbjerg 88

Stochastic Differential Equations
Dietrich E. Wolf 104

Frustrated Systems: Ground State Properties via Combinatorial Optimization
Heiko Rieger 122

Molecular Dynamics
Hans Herrmann 159

Contributors

FLYVBJERG, HENRIK   henrik.flyvbjerg@risoe.dk
   HLRZ c/o Forschungszentrum Jülich, D-52425 Jülich, Germany
   Department of Optics and Fluid Dynamics Risø National Laboratory, DK-
   4000 Roskilde, Denmark.
HERRMANN, HANS   hans@ica1.uni-stuttgart.de
   ICA1, Universität Stuttgart, D-70569 Germany.
KRAUTH, WERNER   krauth@physique.ens.fr
   CNRS-Laboratoire de Physique Statistique, Ecole Normale Supérieure,
   F-75231 Paris Cedex 05, France.
MARINARI, ENZO   marinari@chimera.roma1.infn.it
   Dipartimento di Fisica, Università di Cagliari, via Ospedale 72, 09100 Caglia-
   ri (Italy).
NIEDERMAYER, FERENC   niederma@butp.unibe.ch
   Institute for Theoretical Physics, University of Bern, Sidlerstrasse 5,
   CH-3012 Bern.
RIEGER, HEIKO   heiko@fz-juelich.de
   HLRZ c/o Forschungszentrum Jülich, D-52425 Jülich, Germany.
STAUFFER, DIETRICH   stauffer@thp.uni-koeln.de
   Institute for Theoretical Physics, Cologne University, D-50923 Köln, Ger-
   many.
WOLF, DIETRICH E.   wolf@comphys.Uni-Duisburg.de
   Gerhard-Mercator-Universität, FB 10, D-47048 Duisburg, Germany.

Introduction to Monte Carlo Algorithms

Werner Krauth

CNRS-Laboratoire de Physique Statistique,
Ecole Normale Supérieure, F-75231 Paris Cedex 05, France

Abstract. These lectures that I gave in the summer of 1996 at the Beg-Rohu (France) and Budapest summer schools discuss the fundamental principles of thermodynamic and dynamic Monte Carlo methods in a simple and light-weight fashion. The keywords are *Markov chains, sampling, detailed balance, a priori probabilities, rejections, ergodicity, "Faster than the clock algorithms"*.

The emphasis is on orientation, which is difficult to obtain (all the mathematics being simple). A firm sense of orientation is essential, because it is easy to lose direction, especially when you venture to leave the well trodden paths established by common usage.

The discussion will remain quite basic (and I hope, readable), but I will make every effort to drive home the essential messages: the crystal-clearness of detailed balance, the main problem with Markov chains, the large extent of algorithmic freedom, both in thermodynamic and dynamic Monte Carlo, and the fundamental differences between the two problems.

1 Equilibrium Monte Carlo Methods

1.1 A game in Monaco

The word "Monte Carlo method" can be traced back to a game very popular in Monaco. It's not what you might at first think, but a children's pastime played on the beaches. Each Wednesday (when there is no school) and on the weekends, they get together and draw a circle and a square in the sand with a big stick as shown in Fig. 1. They fill their pockets with pebbles[1], stand around the square, and start throwing pebbles randomly at it with their eyes shut. Someone keeps track of how many pebbles hit the square, and how many landed inside the circle (see Fig. 1). You will easily verify that the ratio of pebbles inside the circle to the pebbles within the whole of the square should come out to be $\pi/4$, so there is much excitement when the 40th, 400th, 4000th pebble is about to be cast.

This breath-taking game is the only method I know for computing the number π to arbitrary precision without using fancy measuring devices (meter stick, balance) or advanced mathematics (division, multiplication, trigonometry). Children in Monaco can pass the whole afternoon at this game. You are invited[2] to write a little program to simulate the game. If you have never written a

---

[1] pebble is *calculus* in Latin.
[2] cf. ref. [1] for an unsurpassed discussion of random numbers, including all practical aspects.

2   Werner Krauth

Fig. 1. Children at play on the beaches of Monaco. They spend their afternoons calculating π by a method which can be easily extended to general integrals.

Monte Carlo program before, this will be your first one. You may also recover "`simplepi.f`" from my WWW-site.

Lately, adults in Monaco have been getting into a similar game. Late in the evenings, when all the helicopters are safely parked, they get together at the

Fig. 2. Adults at play at the Monte Carlo heliport. Their method to calculate π has been extended to the investigation of liquids, suspensions, and lattice gauge theory.

local heliport (cf. Fig. 2), which offers the same basic layout as in the children's game. They fill their expensive Hermès handbags with pebbles, but, since the field is so large, they play a slighty different game: they start somewhere on the field, close their eyes, and then throw the little stone in a random direction. Then they walk to where the stone has landed, take a new one out of their handbag, and start again. You can see that using this method, one can also sweep out the heliport square evenly, and compute the number π. You are invited to write a 10-line program to simulate the heliport game - but it's not completely trivial.

Are you starting to think that the discussion is too simple? If so, please consider the lady at c). She has just flung a pebble to c'), which is outside the square. What should she do?

1. throw another pebble from c)
2. climb over the fence, and continue until, by accident, she gets back to the heliport
3. other: [ ]

A satisfactory answer to this question will be given on p.8: it contains the essence of the concept of detailed balance. Many Monte Carlo programs are faulty because the wrong alternative was programmed.

The two cases - the children's game and the grown-ups' game - are perfect illustrations of what is going on in Monte Carlo... and in Monte Carlo algorithms. In each case, one is interested in evaluating an integral

$$\int_{x,y\in\square} dxdy\,\pi(x,y)f(x,y), \tag{1}$$

with a probability density π which, in our case is the square

$$\pi(x,y) = \begin{cases} 1 \text{ if } |x| < 1 \text{ and } |y| < 1 \\ 0 \text{ otherwise} \end{cases} \tag{2}$$

and a function f (the circle)

$$f(x,y) = \begin{cases} 1 \text{ if } x^2 + y^2 < 1 \\ 0 \text{ otherwise} \end{cases}. \tag{3}$$

Both the children and the grown-ups fill the square with a constant density of pebbles (corresponding to $\pi(x,y) = 1$.), one says that they *sample* the function $\pi(x,y)$ on the basic square. If you think about it you will realize that the number π can be computed in the two games only because the area of the basic square is known. If this was not so, one would be reduced to computing the ratio of the areas of the circle and of the square, i.e., in the general case, the ratio of the integrals:

$$\int_{x,y\in\square} dxdy\,\pi(x,y)f(x,y) \bigg/ \int_{x,y\in\square} dxdy\,\pi(x,y). \tag{4}$$

Two basic approaches are used in the Monte Carlo method:
1. direct sampling (children on the beach)
2. Markov-chain sampling (adults at the heliport)

Direct sampling is usually like pure gold, because it means that you can call a subroutine which provides an independent hit at your distribution function $\pi(x, y)$. This is exacty what the kids do whenever they get a new pebble out of their pockets.

In most cases no (reasonable) direct sampling algorithm is available. One then resorts to Markov-chain sampling, like the adults at the heliport. Almost all physical systems belong to this class. A famous example, which does not oblige us to speak of energies, Hamiltonians etc. has occupied more than a generation of physicists. It can easily be modelled with a shoe-box, and a number of coins (cf. Fig. 3): how do you generate (directly sample) random configurations of the coins

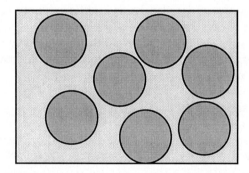

Fig. 3. The Coin in a shoe-box problem (hard disks) has occupied Monte Carlo workers since 1953. There is no direct-sampling algorithm, and Markov-chain sampling is extremely tedious at some densities

such that they don't overlap? Imagine the $2N$-dimensional configuration space of N non-overlapping coins in a shoe-box. Nobody has found a subroutine which would directly sample this configuration space, i.e. create any of its members with equal probability. People's first reaction to this problem is often to propose a method called *Random Sequential Adsorption*: deposit the first coin at a random position, then the second, etc., (if they overlap, simply try again). Random Sequential Adsorption will be dealt with in detail, but in a completely different context, on page 30. Try to understand that this has nothing to do with finding a random non-overlapping configuration of the coins in the shoe-box (in particular, the maximum density of random sequential deposition is much smaller than the close packing density of the coins).

Direct sampling is usually impossible - and that is the basic problem of Monte Carlo methods (in statistical physics). In contrast, the grown-ups' game can be played in any situation (...for example on the heliport which is too large for direct sampling). Markov chain sampling has been used in an uncountable number of statistical physics models: in the aforementioned coin-in-a-box problem, the Hubbard model, etc. In general it is a very poor, and extremely costly substitute for what we really want to do.

In the introduction, I was talking about orientation, and to get oriented you should realize the following:

- In the case of the children's game, you need only a few dozen pebbles (samples) to get a physicist's idea of the value of π, which will be sufficient for most applications. Exactly the same is true for some of the most difficult statistical physics problems. A few dozen (direct) samples of large coin-in-the-box problems at any density would be sufficient to resolve long-standing controversies. (For a random example, look up ref. [2], where the analysis of a monumental simulation relies on an effective number of $4 - 5$ samples). Similarly, a few dozen direct samples of the Hubbard model of strong fermionic correlation would give us important insights into superconductivity. Yet, some of the largest computers in the world are churning away day after day on problems similar to the ones mentioned. They are all trying to bridge the gap between the billions of Markov chain samples and the equivalent of a few random flings in the children's game.
- It is only after we have understood the problem of generating independent samples by Markov chains, that we can start worrying about the slow convergence of mean-values. This is already apparent in the children's game - as in every measurement in experimental physics - : the precision of the numerical value decreases only as $1/\sqrt{N}$, where N is the number of independent measurements. Again, let me argue against the "mainstream" by saying that absolute precision is not nearly as important as it may seem: you are not always interested in computing your specific heat to five significant digits before you can put your soul to rest. In the daily practice of Monte Carlo it is usually more crucial to be absolutely sure that the program has given some independent samples[3] than that there are millions and billions of them.
- It is essential to understand that a long Markov chain, even if it produces only a small number of independent samples (and thus a very approximative result) is usually extremely sensitive to bugs, and even small irregularities of the program.

1.2 The toddlers' algorithm and detailed balance

We have yet to determine what the lady at point c) in the heliport game should do. It's a difficult question, not without consequences, and we don't want to give her any false advice. So, let's think, and first analyze a similar, discretized game, the well-known puzzle shown in Fig. 4, before making a definite recommendation. The task now is to create a perfect scrambling algorithm which generates all

---

[3] i.e. that it has decorrelated from the initial configuration

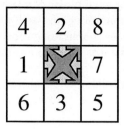

Fig. 4. This well-known puzzle is the starting point for our theoretical analysis of the Monte Carlo method.

possible configurations of the puzzle with equal probability. One such well-known method, the toddlers' algorithm, is illustrated in Fig. 5 together with one of its inventors.

Fig. 5. This is a direct sampling algorithm for the puzzle game.

The theoretical baggage picked up in the last few sections allows us to class this method without hesitation among the direct sampling methods (as the children's game before), since it creates an independent instance each time the puzzle is taken apart.

We would be more interested in a grown-up people's algorithm, which respects the architecture of the game[4].

What would you do? Almost certainly, you would hold the puzzle in your

---

[4] cf. the discussion on page 21 for an important and subtle difference between the toddlers' algorithm and any incremental method.

hands and - at each time step - move the empty square in a random direction. The detailed balance condition which you will find out about in this section shows that this completely natural sounding algorithm is wrong.

If the blank square is in the interior (as in Fig. 4), then clearly the algorithm should move the empty square in one of the four possible directions ↑ → ↓ ← with equal probability.

As in the heliport game, we have to analyze what happens at the boundary of the square[5]. Consider the corner configuration a, which communicates with

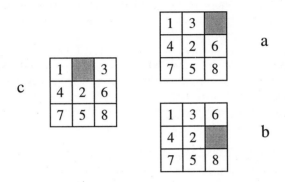

Fig. 6. The corner configuration a communicates with configurations b and c

the configurations b and c, as shown in Fig. 6. If our algorithm (yet to be found) generates configurations a, b and c with probabilities $\pi(a)$, $\pi(b)$, and $\pi(c)$, respectively (we require them to be the same), we can derive a simple rate equation relating the π's to the transition probabilities $p(a \to b)$ etc. Since a can only be generated from b, c, or from itself, we have

$$\pi(a) = \pi(b)p(b \to a) + \pi(c)p(c \to a) + \pi(a)p(a \to a), \tag{5}$$

this gives

$$\pi(a)[1 - p(a \to a)] = \pi(b)p(b \to a) + \pi(c)p(c \to a). \tag{6}$$

The condition that tells us that the empty square, once at a, can stay at a or move to b or c will be:

$$1 = p(a \to a) + p(a \to b) + p(a \to c), \tag{7}$$

which gives $[1 - p(a \to a)] = p(a \to b) + p(a \to c)$. This formula, introduced in (6), yields

$$\pi(a)\underbrace{p(a \to b) + \pi(a)p(a \to c)} = \overbrace{\pi(c)p(c \to a) + \pi(b)}p(b \to a). \tag{8}$$

---

[5] Periodic boundary conditions are *not* allowed.

We can certainly satisfy this equation if we equate the braced terms separately:

$$\pi(a)p(a \to b) = \pi(b)p(b \to a) \quad \text{etc..} \tag{9}$$

This equation is the celebrated condition of detailed balance.

We admire it for a while, and then pass on to its first application, in our puzzle. There, we impose equal probabilities for all accessible configurations, i.e. $p(a \to b) = p(b \to a)$, etc, and the only simple way to connect the proposed algorithm for the interior of the square with the boundary is to postulate an equal probability of $1/4$ for any possible move. Looking at (7), we see that we have to allow a probability of $1/2$ for $p(a \to a)$. The consequence of this analysis is that to maximally scramble the puzzle, we have to be immobile with a probalility $1/2$ in the corners, and with probability $1/4$ on the edges.

We can assemble all the different rules in the following algorithm: At each time step $t = 0, 1, \ldots$

1. choose one of the four directions $\uparrow \quad \to \quad \downarrow \quad \leftarrow$ with equal probabilities.
2. move the blank square in the corresponding direction if possible. Otherwise, stay where you are, but advance the clock.

I have presented this puzzle algorithm hoping that you will *not* believe that it works. This may take you to scrutinize the detailed balance condition, and to understand it better.

If you recall that the puzzle game was in some respects a discretized version of the heliport game, you will now be able to resolve the fashionable lady's dilemma by analogy. Notice that there is already a stone on the ground where the lady is standing. She should now pick two more pebbles out of her handbag, place one on top of the one already there, and use the other one to try a new fling. If this is again an out-fielder, she should pile one more stone up etc.. If you look at the heliport after the game has been going for a while, you will notice a strange pattern of pebbles on the ground (cf. Fig. 7). There are mostly single stones in the interior, and some piles of varying heights as you approach the boundary, especially the corner. Yet, this is the standard way to enforce a constant density $\pi(x, y) = const$.

People hearing this for the first time are sometimes struck with utter incredulity. They find the piling-up of stones absurd and conclude that *I* must have got the story wrong. The only way to reinstall confidence is to show simulations (of ladies flinging stones) which do and do not pile up stones on occasions. You are invited to do the same (start with one dimension).

In fact, we have just seen the first instance of a *rejection*, which, as announced, is a keyword of this course, and of the Monte Carlo method in general. The concept of a rejection is so fundamental that it is worth discussing it in the completely barren context of a Monaco airfield. Rejections are the basic method by which Monte Carlo enforces a correct density $\pi(x, y)$ on the square, with an algorithm (random "flings") which is not particularly suited to the geometry. Rejections are also wasteful (of pebbles), and expensive (you throw stones but

Fig. 7. Pebbles at the end of the heliport game. Notice the piles at the edges, particularly close to the corners.

in the end you just stay where you were). We will deepen our understanding of rejections considerably in the next sections.

We have introduced rejections by a direct inspection of the detailed balance condition. This trick has been elevated to the status of a general method in the so-called Metropolis algorithm. There, the detailed balance condition (9) is satisfied by using

$$P(a \to b) = \min(1, \frac{\pi(b)}{\pi(a)}). \qquad (10)$$

What does this equation mean? Let us find out in the case of the heliport: standing at a) (which is inside the square, i.e. $\pi(a) = 1$), you throw your pebble in a random direction (to b). Two things can happen: Either b is inside the square ($\pi(b) = 1$), and (10) tells you to accept the move with probability 1, or b is outside ($\pi(b) = 0$), and you should accept with probability 0, i.e. reject the move, and stay where you are.

After having obtained an operational understanding of the Metropolis algorithm, you may want to see whether it works in the general case. For once, there is rescue through bureaucracy, for the theorem can be checked by a bureaucratic procedure: simply fill out the following form:

case	$\pi(a) > \pi(b)$	$\pi(b) > \pi(a)$	
$P(a \to b)$			1
$\pi(a) P(a \to b)$			2
$P(b \to a)$			3
$\pi(b) P(b \to a)$			4

Fill it out, and you will realize that in both columns the second and forth rows are identical, as stipulated by the detailed balance condition. You have understood all there is to the Metropolis algorithm.

1.3 To cry or to Cray

Like a diamond tool, the Metropolis algorithm allows you to "grind down" an arbitrary density of trial movements (as the random stone-flings) into the chosen stationary probability density $\pi(x, y)$.

To change setting we now discuss how a general model from classical statistical physics with an arbitrarily high-dimensional energy $E(x_1, x_2, \ldots, x_N)$ is simulated. In this case, the probability density is given by the Boltzmann measure $\pi(x_1, ..., x_N) = \exp(-\beta E)/Z$, where β is the inverse temperature, Z the partition function and the physical expectation values (energies, expectation values) are given by formulae such as (4). All you have to do (if the problem is not too difficult) is to ...

- Set up a (rather arbitrary) displacement rule, which should generalize the random stone-flings of the heliport game. For example, you may go from an initial configuration $x_1, x_2, \ldots, x_N$ to a new configuration by choosing an arbitrary dimension i, and doing the random displacement on $x_i \to x_i + \delta x$, with δx between $-\epsilon$ and $+\epsilon$. Discrete variables are treated just as simply.
- Having recorded the energies E^a and E^b at the initial point a and the final point b, you may use the Metropolis algorithm to compute the probability p, to actually get there:

$$p(a \to b) = \min[1, \exp(-\beta(E^b - E^a))]. \qquad (11)$$

This is implemented using a single uniformly distributed random number $0 < ran < 1$, and we move our system to b under the condition that $ran < p(a \to b)$, as shown in the figure.

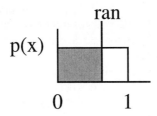

Fig. 8. The probability for a uniformly distributed random number $0 < ran < 1$ to be smaller than $p(a \to b)$ is $\min[1, p(a \to b)]$. A typical Monte Carlo program contains one such comparison per step.

- You can notice that (for continuous systems) the remaining liberty in your algorithm is to adjust the value of ϵ. The time-honored procedure is to choose ϵ such that about half of the moves are rejected. This is of course just a golden rule of thumb. The "$< p(\epsilon) > \sim 1/2$" rule, as it may be called, *does* in general insure quite a quick diffusion of your particle across configuration space. If you want to do better, you have to monitor the speed of your Monte Carlo diffusion, but it is usually not worth the effort.

As presented, the Metropolis algorithm is extremely powerful, and many problems are adequately treated with this method. The method is so powerful that for many people the theory of Monte Carlo stops right after equation (9). All the rest are implementation details, data collection, and the adjustment of ϵ, which was just discussed.

Many problems, especially in high dimensions, defy this rule, however. For these problems, programs written along the lines of the one presented above will run properly, but will have a very hard time generating independent samples. These are the problems which force one to either give up or compromise: use smaller sizes than one wants, take risks with the error analysis etc.. The published papers, over time, present much controversy, shifting results, and much frustration to the student executing the calculation.

Prominent examples of difficult problems are phase transitions in general statistical models, the Hubbard model, bosons, and disordered systems. The strong sense of frustration can best be retraced in the case of the hard sphere liquid, which was first treated in the original 1953 publication introducing the Metropolis algorithm, and which has since generated an unabating series of unconverging calculations, and heated controversies.

A very useful system to illustrate a difficult simulation, shown in Fig. 9 is

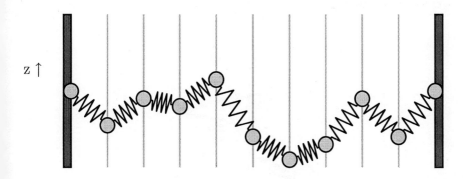

Fig. 9. Chain of coupled springs which serves as an example of a large-dimensional Monte Carlo problem.

the chain of N springs with energy

$$E = \frac{(z_1 - z_0)^2}{2} + \frac{(z_2 - z_1)^2}{2} + \ldots + \frac{(z_{N+1} - z_N)^2}{2}, \quad (12)$$

z_0 and z_{N+1} are fixed at zero, and $z_1, \ldots z_N$ are the lateral positions of the points, the variables of the system. How do we generate distributions according to $\exp(-\beta E)$? A simple algorithm is very easily written (you may recover "spring.f" from my WWW-site). According to the recipe given above you simply choose a random bead and displace it by a small amount. The whole program can be written in about 10 lines. If you experiment with the program, you can also try to optimize the value of ϵ (which will turn out to be optimal when the average change in energy in a single move is $\beta|E^b - E^a| \sim 1$).

What you will find is that the program works, but is extremely slow. It is so slow that you want to cry or to *Cray* (give up or use a supercomputer), and both would be ridiculous.

Let us analyze why a simple program for the coupled springs is so slow. The reason are the following

- Your "$< p(\epsilon) > \sim 1/2$" rule fixes the step size, which is necessarily very small.
- The distribution of, say, $z_{N/2}$, the middle bead, as expressed *in units of ϵ* is very large, since the whole chain will have a lateral extension of the order of $\sqrt{N}$. It is absolutely essential to realize that the distribution can never be intrinsically wide, but only in units of the imposed step size (which is a property of the algorithm).
- It is very difficult to sample a large distribution with a small stepsize, in the same way as it takes a very large bag of stones to get an approximate idea of the numerical value of π if the heliport is much larger than your throwing range (cf. Fig. 10).

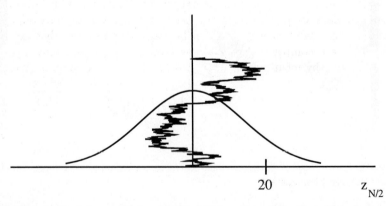

Fig. 10. The elastic string problem with $N = 40$ is quite a difficult problem because the distribution of, e.g. $z_{N/2}$ is large. The simulation path shown in the figure corresponds to 400.000 steps.

The difficulty in sampling a wide distribution is the basic reason why simulations can have difficulties converging.

At this point, I often encounter the remark: why can't I move several interconnected beads independently, thus realizing a large move? This strategy is useless. In the big people's game, it corresponds to trying to save every second stone by throwing it into a random direction, fetching it (possibly climbing over the fence), picking it up and throwing it again. You don't gain anything since you have already optimized the throwing range before. You already had a large probability of rejection, which will now become prohibitive. The increase in the rejection probability will more than annihilate the gain in stride.

The way it is set up, the thermal spring problem is difficult because the many degrees of freedom $x_1, \ldots x_N$ are strongly coupled. Random ϵ-moves are an extremely poor way to deal with such systems. Without an improved strategy for the attempted moves, the program very quickly fails to converge, i.e. to produce even a handful of independent samples.

In the coupled spring problem, there are essentially two ways to improve the situation. The first is to use a basis transformation, which in this case is simply to go to Fourier space. This evidently decouples the degrees of freedom. You may identify the Fourier modes that have a chance of being excited. If you write a little program, you will very quickly master a popular concept called "Fourier acceleration". An exercise of unsurpassed value is to extend the program to an energy which contains a small additional anisotropic coupling of the springs and treat it with both algorithms. Fermi, Pasta and Ulam, in 1945, were the first to simulate the anisotropic coupled chain problem on a computer (with a deterministic algorithm) and to discover extremely slow thermalization.

The basis transformation is a specific method to allow large moves. Usually however, it is not possible to find such a transformation. A more general method to deal with difficult interacting problems consists of isolating subproblems which you can sample more or less easily and solve exactly. The *a priori* information gained from this analytical work can then be used to propose exactly those (large) moves which are compatible with the system. The proposed moves are then rendered correct by means of a generalized Monte Carlo algorithm. A modified Metropolis rejection remains, correcting the "engineered" density into the true stationary probability. We will first motivate this very important method in the case of the coupled chain example, and then give the general theory, and present a very important application to spin models.

To really understand what is going on in the coupled spring problem, let's go back to Fig. 9, and analyze only a part of the whole game: the motion of the bead labelled i with $i-1$ and $i+1$ (for the moment) *immobilized* at some values z_{i-1} and at z_{i+1}. It is easy to do this simulation (as a thought experiment) and to see that the results obtained are as given in Fig. 11. You can see that the distribution function $P(z_i)$, (at fixed z_{i-1} and z_{i+1}) is a Gaussian centered around $\bar{z}_i = (z_{i-1} + z_{i+1})/2$. Note, however, that there are in fact two distributions: the accepted one and the rejected one. It is the Metropolis algorithm which, through the rejections, modifies the proposed - flat - distribution into the

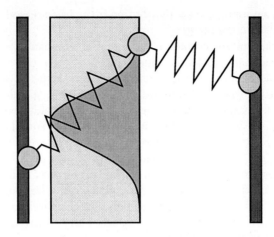

Fig. 11. Here we analyze part of the coupled spring simulation, with z_{i-1} and z_{i+1} immobilized. The large rejection rate is due to the fact that the proposed probability distribution for z_i (light gray) is very different from the accepted distribution (the Gaussian).

correct Gaussian. We see that the rejections - besides being a nuisance - play the very constructive role of modifying the proposed moves into the correct probability density. There is a whole research literature on the use of rejection methods to sample 1−dimensional distributions (cf. ref [1] chap 7.3), ... a subject which we will leave instantly because we are more interested in the higher-dimensional case.

1.4 A priori probability

Let us therefore extend the usual formula for the detailed balance condition and for the Metropolis algorithm by taking into account the possible "*a priori*" choices of the moves, which are described by an *a priori probability* $\mathcal{A}(a \to b)$ to attempt the move. In the heliport game, this probability was simply

$$\mathcal{A}(a \to b) = \begin{cases} 1 \text{ if } |a - b| < \epsilon \\ 0 \text{ otherwise} \end{cases}, \tag{13}$$

with ϵ being the throwing range, and we did not even notice its presence. In the elastic spring example, the probability to pick a bead i, and to move it by a small amount $-\epsilon < \delta < \epsilon$ was also independent of i, and of the position z_i at the given moment.

Now, we reevaluate the detailed balance equation, and allow explicitly for an algorithm: The probability $p(a \to b)$ is split into two separate parts [3]:

$$p(a \to b) = \mathcal{A}(a \to b)\mathcal{P}(a \to b), \tag{14}$$

where $\mathcal{P}(a \to b)$ is the (still necessary) acceptance probability of the move proposed with $\mathcal{A}(a \to b)$. What is this rejection probability? This is very easy to obtain from the full detailed balance equation

$$\pi(a)\mathcal{A}(a \to b)\mathcal{P}(a \to b) = \pi(b)\mathcal{A}(b \to a)\mathcal{P}(b \to a). \tag{15}$$

You can see that for any *a priori* probability, i.e. for *any* Monte Carlo algorithm we can find the acceptance rate which is needed to bring this probability into accordance with our unaltered detailed balance condition. As before, we can use a Metropolis algorithm to obtain (one possible) correct acceptance rate

$$\frac{\mathcal{P}(a \to b)}{\mathcal{P}(b \to a)} = \frac{\pi(b)}{\mathcal{A}(a \to b)} \frac{\mathcal{A}(b \to a)}{\pi(a)}, \tag{16}$$

which results in

$$\mathcal{P}(a \to b) = \min\left(1, \frac{\pi(b)}{\mathcal{A}(a \to b)} \frac{\mathcal{A}(b \to a)}{\pi(a)}\right). \tag{17}$$

Evidently, this formula reduces to the original Metropolis prescription if we introduce a flat *a priori* probability (as in (13)). As it stands, (17) states the basic algorithmic liberty which we have in setting up our Monte Carlo algorithms: any possible bias $\mathcal{A}(a \to b)$ can be made into a valid method, since we can always correct it with the corresponding acceptance rate $\mathcal{P}(a \to b)$. Of course, only a very carefully chosen probability will be viable, or even superior to the simple and popular choice, (10).

Inclusion of a general *a priori* probability is mathematically harmless, but generates a profound change in the practical setup of our simulation. In order to evaluate the acceptance probability $\mathcal{P}(a \to b)$ in (17), we not only propose the move to b, but also need explicit evaluations of both $\mathcal{A}(a \to b)$ *and* of the return move $\mathcal{A}(b \to a)$.

Can you see what is new in (17) compared to the earlier ((10))? Before, we necessarily had a large rejection probability whenever we were going from a point a with high probability (large $\pi(a)$) to a point b which had a low probability (small $\pi(b)$). The naive Metropolis algorithm could only produce the correct probability density by installing rejections. *Now* we can counteract, by simply choosing an *a priori* probability which is also much smaller. In fact, you can easily see that there is an an optimal choice: we may be able to use as an *a priori* probability $\mathcal{A}(a \to b)$ the probability density $\pi(b)$ and for $\mathcal{A}(b \to a)$ the probability density $\pi(a)$. In that case, the ratio expressed in (17) will always be equal to 1, and there will never be any rejections. Of course, we are then also back to direct sampling ... from where we came from because direct sampling was too difficult ....

The argument is not circular, as it may appear, because it can always be applied to a part of the system only. To understand this point, it is best to go back to the example of the elastic string. We know that the probability distribution $\pi(z_i)$ for fixed z_{i-1} and z_{i+1} is

$$\pi(z_i | z_{i-1}, z_{i+1}) \sim \exp[-\beta(z_i - \bar{z}_i)^2], \tag{18}$$

with $\bar{z}_i = (z_{i-1} + z_{i+1})/2$. We can now use exactly this formula as an *a priori* probability $\mathcal{A}(z_i|z_{i-1}, z_{i+1})$ and generate an algorithm without rejections, which thermalizes the bead i at any step with its immediate environment. To program this rule, you need Gaussian random numbers (cf. [1] for the popular Box-Muller algorithm). So far, however, the benefit of our operation is essentially non-existent[6].

It is now possible to extend the formula for z_i at fixed z_{i-1} and z_{i+1} to a larger window. Integrating over z_{i-1} and z_{i+1} in $\pi(z_{i-2}, \ldots, z_{i+2})$, we find that

$$\pi(z_i|z_{i-1}, z_{i+1}) \sim \exp[-2\beta(z_i - \bar{z}_i)^2], \qquad (19)$$

where now $\bar{z}_i = (z_{i-2} + z_{i+2})/2$. Having determined z_i from a sampling of (19), we can subsequently find values for z_{i-1} and z_{i+1} using (18). The net result of this is that we can update z_{i-1}, z_i, z_{i+1} simultaneously. The program "`levy.f`" which implements this so called Lévy construction can be retrieved and studied from my WWW-site. It generates large moves with an arbitrary window size without rejections.

1.5 Perturbations

From this simple example of the coupled spring problem, you can quickly reach all the subtleties of the Monte Carlo method. You can see that we were able to produce a perfect algorithm, because the *a priori* probability $\mathcal{A}(z_i|z_{i-l}, z_{i+l})$ could be chosen to equal the stationary probability $\pi(z_{i-l}, \ldots, z_{i+l})$, resulting in a vanishing rejection rate. This, however, was just a happy accident. The enormous potential of the *a priori* probability resides in the fact that (17) (usually) deteriorates very gracefully when $\mathcal{A}$ and $\pi(z)$ differ. A recommended way to understand this point is to write a second program for the coupled chain problem, in which you again add a little perturbing term to the energy, such as

$$E_1 = \gamma \sum_{i=1}^{N} f(z_i), \qquad (20)$$

which is supposed to be relatively less important than the elastic energy. It is interesting to see in what way the method must be modified if we keep the above Lévy-construction as our algorithm[7]. If you go through the following argument, and possibly even write the program and experiment with its results, you will find the following

- The energy of each configuration is now $\tilde{E}(z_1, \ldots, z_N) = E_0 + E_1$, where $E_0(a)$ is the term given in (12), which is in some way neutralized by the *a*

---

[6] The algorithm with this *a priori* probability is called "heatbath" algorithm. It is popular in spin models, but essentially identical to the Metropolis method.

[7] In quantum Monte Carlo you introduce a small coupling between several of the strings

priori probability $\mathcal{A}(a \to b) = \exp[-\beta E_0(b)]$. One can now write down the Metropolis acceptation rate of the process from (17). The result is

$$\mathcal{P}(a \to b) = \min\left(1, \frac{\exp[-\beta E_1(b)]}{\exp[-\beta E_1(a)]}\right). \tag{21}$$

This is exactly as in the naive Metropolis algorithm, (10), but exclusively with the newly added term of the energy.

- Implementing the *a priori* probability with $\gamma = 0$, your code runs with acceptance probability 1, independent of the size of the interval $2l+1$. If you include the second term E_1, you will again have to optimize this size. Of course, you will find that the optimal window corresponds to a typical size of $\beta|E_1^b - E_1^a| \sim 1$.
- With this cutting-up of the system into a part which you can solve exactly and an additional term, you will find that the Monte Carlo algorithm has the appearance of a perturbation method. Of course, it will always be exact. It has best chances of being fast if E_1 is generally smaller than E_0. One principal difference to perturbation methods is that it will always sample the full perturbed distribution $\pi(z_1, \ldots, z_N) = \exp[-\beta \tilde{E}]$.

One can spend an essentially endless amount of time pondering about the a priori probabilities, and the similiarities and differences to perturbation theory. This is where the true power of the Monte Carlo method lies. This is what one has to understand before venturing into mass tailoring tomorrow's powerful algorithms to problems which are today thought to be out of reach. Programming a simple algorithm for the coupled oscillators will be an excellent introduction to the subject. Useful further literature are [3], where the coupled spring problem is extended into one of the most successful applications to quantum Monte Carlo methods, and [4], where some limitations of the approach are outlined.

In any case, you should understand that a large rate of rejections is always indicative of a structure of the proposed moves which is badly suited to the probability density of the model at the given temperature. The benefit of fixing this problem - if we can only see how to do it - is usually tremendous: doing the small movements with negligible rejection rates often allows us to do larger movements and to explore the important regions of configuration space all the more quickly.

To end this section, I will give another example of the celebrated cluster methods in lattice systems, which were introduced ten years ago by Swendsen and Wang [5] and by Wolff [6]. There again, we find the two essential ingredients of slow algorithms: necessarily small moves, and a wide distribution of the physical quantities. Using the concept of *a priori* probabilities, we can very easily understand these methods.

1.6 Cluster algorithms

I will discuss these methods in a way which will clearly bring out the "engineering" aspect of the *a priori* probability, as one tries to cope with a large distribution problem. Before doing this, let us discuss, as before, the general setting and the physical origin of slow convergence. As we all know, the Ising model is defined as a system of spins on a lattice with an energy of

$$E = - \sum_{<i,j>} S_i S_j, \qquad (22)$$

where the spins can take on values of $S_i = \pm 1$, and where the sum runs over pairs of neighboring lattice sites. A simple program is again written in a few lines: it picks a spin at random, and tries to turn it over. Again, the a priori probability is flat, since the spin to be flipped is chosen arbitrarily. You may find such a program ("simpleising.f") on my WWW site. Using this program, you can easily recover the phase transition between a paramagnetic and a ferromagnetic phase, which in two dimensions takes place at a temperature of $\beta = \log(1+\sqrt{2})/2$ (you may want to look up the classic reference [7] for exact results on finite lattices). You will also find that the program is increasingly slow around the critical temperature. This is usually interpreted as the effect of the divergence of the correlation length as we approach the critical point. In our terms, we understand this slowing down equally well: our simple algorithm changes the magnetization by at most a value of 2 at each step, since the algorithm flips only single spins. This discrete value replaces the ϵ in our previous example of the coupled springs. If we now plot histograms of the total magnetization of the system (in units of the step size $\Delta m = 2$!), we again see that this distribution becomes "wide" as we approach β_c. (cf. Fig. 12) Clearly, the total magnetization

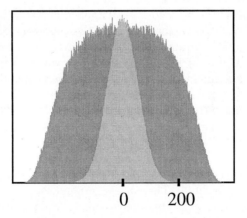

Fig. 12. Histograms of the magnetization in the 20×20 Ising model at temperatures 10 % and 45 % above the critical point. On the x-axis is plotted $m/\Delta m$, where $\Delta m = 2$ is the change of the magnetization which can be obtained in a single Monte Carlo step.

has a wide distribution, that is extremely difficult to sample with a single spin-flip algorithm.

Introduction to Monte Carlo Algorithms 19

To appreciate cluster algorithms, you have to understand two things:

1. As in the coupled spring problem, you cannot simply flip several spins simultaneously (cf. the discussion on page 13.) You want to flip large clusters, but you cannot of course simply solder together all the spis of one sign which are connected to each other, because those could never again be separated.
2. If you cannot solidly fix the spins of same sign with probability 1, you have to choose adjacent coaligned spins with a certain probability p. This probability p is the construction parameter of our a priori probability $\mathcal{A}$. The algorithm will run for an arbitrary value of p ($p = 0$ corresponding to the single spin-flip algorithm), but the p minimizing the rejections will be optimal.

The cluster algorithm we find starts with the idea of choosing an arbitrary starting point, and adding "like" links with probability p. We arrive here at the first nontrivial example of the evaluation of an a priori probability. Imagine that we start from a "+" spin in the gray area of configuration a) and add like spins. What is the a priori probability $\mathcal{A}(a \to b)$ and the inverse probability $\mathcal{A}(b \to a)$, and what are the Boltzmann weights $\pi(a)$ and $\pi(b)$?

$\mathcal{A}(a \to b)$ is given by a term involving interior "+ +" links, $\mathcal{A}_{int}(a \to b)$, which looks difficult, and which we won't even try to evaluate, and a part concerning the boundary of the cluster. This boundary is made up of two types of links, as summarized below:

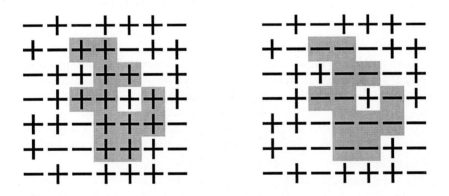

Fig. 13. construction process of the Swendsen-Wang algorithm. Starting from an initial + site, we include other + sites with probability p (left). The whole cluster (gray area) is then flipped. In the text, we compute the probability of stopping the construction of the gray cluster, and of the reverse move. This yields our a priori probabilities.

int	ext	number
+	−	c_1
+	+	c_2

$$E|_{\partial C} = -c_2 + c_1 \qquad (23)$$

(in the example of Fig. 13, we have $c_1 = 10$ and $c_2 = 14$). The *a priori* probability is $\mathcal{A}(a \to b) = \mathcal{A}_{int} \times (1-p)^{c_2}$. To evaluate the Boltzmann weight, we also abstain from evaluating terms which don't change between a) and b): clearly we only need the boundary energy, which is given in (23). It follows that $\pi(a) \sim \exp[-\beta(c_1 - c_2)]$. We now consider the inverse move. In cluster b), the links across the boundary are as follows:

int	ext	number
−	−	c_1
−	+	c_2

$$E|_{\partial C} = -c_1 + c_2. \qquad (24)$$

The *a priori* probability to construct this cluster is again put together from an interior part, which is exactly the same as for the cluster in a), and a boundary part, which is changed: $\mathcal{A}(b \to a) = \mathcal{A}_{int} \times (1-p)^{c_1}$. Similarly, we find that $\pi(a) \sim \exp[-\beta(c_2 - c_1)]$. We can now put everything into the formula for detailed balance:

$$e^{-\beta[c_1 - c_2]}(1-p)^{c_2} \mathcal{P}(a \to b) = e^{-\beta[c_2 - c_1]}(1-p)^{c_1} \mathcal{P}(b \to a), \qquad (25)$$

which results in the acceptance probability:

$$\mathcal{P}(a \to b) = \min(1, \frac{e^{-\beta[c_2 - c_1]}(1-p)^{c_1}}{e^{-\beta[c_1 - c_2]}(1-p)^{c_2}}. \qquad (26)$$

The most important point about this equation is *not* that it can be simplified, as we will see in a minute, but that it is perfectly explicit, and may be evaluated without trouble: once your cluster is constructed, you could simply evaluate c_1 and c_2 and compute the rejection probability from this equation.

On closer inspection of (26), you can see that, for $1 - p = \exp[-2\beta]$, the acceptance probability is always 1. This is the "magic" value implemented in the algorithms of Swendsen-Wang and of Wolff.

The resulting algorithm [6] is very simple, and closely follows the description given above: you start picking a random spin, and add coaligned neighbors with probability p, the construction stops when none of the "like" links across the growing cluster have been chosen. If you are interested, you may retrieve the program "wolff.f" from my WWW-site. This program (which was written in less than an hour) explains how to keep track of the cluster construction process. It is amazing to see how it passes the Curie point without any perceptible slowing down.

Once you have seen such a program churn away at the difficult problem of a statistical physics model close to the critical point you will come to understand what a great pay-off can be obtained from an intelligent use of powerful Monte Carlo ideas.

1.7 Concluding remarks on equilibrium Monte Carlo

We have arrive to the end of the introduction to equilibrium Monte Carlo methods. I hope to have given a comprehensible introduction to the way the method presents itself in most statistical physics contexts:

- The (equilibrium) Monte Carlo approach is an integration method which converges slowly, but surely. Except for a few cases, one is always forced to sample the stationary density (Boltzmann measure, density matrix in the quantum case) by a Markov chain approach. The critical issue there is the correlation time, which can become astronomical. In a typical application, one is happy with a very small number of truly independent samples, but an appallingly large number of computations will never succeed in decorrelating the Markov chain from the initial condition.
- The standard methods work well, but have some important disadvantages. As presented, the condition that the rejection rate has to be quite small - typically of the order of $1/2$ - reduces us to very local moves.
- The acceptance rate has important consequences for the speed of the program, but a small acceptance rate in particular is an indicator that the proposed moves are inadequate.
- It would be wrong to give the impression that the situation can always be ameliorated - sometimes one is simply forced to do very big calculations. In many cases, however, a judicious choice of the *a priori* probability allows us to obtain very good acceptance rates, even for large movements. This work is *important*, and its pay-off is some sort of exponential.

2 Dynamical Monte Carlo Methods

2.1 Ergodicity

Usually, together with the fundamental concept of detailed balance, one also finds a discussion of *ergodicity*, since it is the combination of both principles that insures that the simulation will converge to the correct probability density. Ergodicity simply means that any configuration b can eventually be reached from the initial configuration a, and we denote it by $p(a \to \ldots \to b) > 0$.

Ergodicity is a tricky concept, which does not have a step-by-step practical meaning like detailed balance. There are two reasons why the ergodicity condition can fail to be satisfied:

- Trivially, your Monte Carlo might for some reason only sample some part of phase space, split off, e.g. by a symmetry principle. An amusing example is given by the puzzle game we considered in section 1.2 : it is easy to convince yourself that the toddler's algorithm is *not* equivalent to the grown-up algorithm, and that it creates only one half of the possible configurations. Consider those configurations of the puzzle where the empty square in a given position, say in the lower right corner, as in Fig. 14. Two such config-

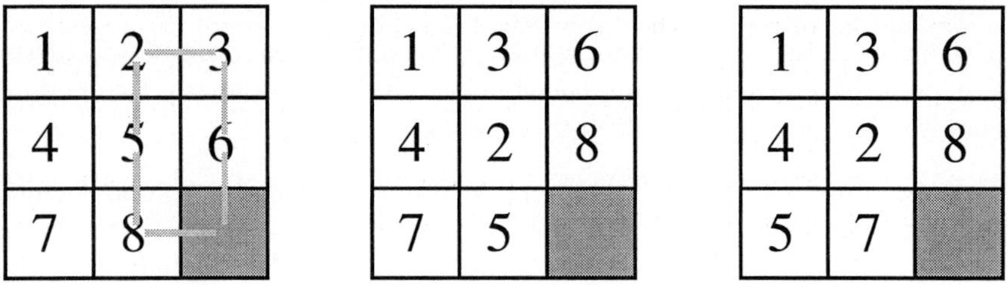

Fig. 14. The two configurations to the left can reach one-another via the path indicated (the arrows indicate the (clockwise) motion of the empty square). The rightmost configuration cannot be reached since it differs by only one transposition from the middle one.

urations can only be obtained from each other if the empty square describes a closed path, and this invariably corresponds to an *even* number of steps (transpositions) of the empty square. The first two configurations in Fig. 14 can be obtained by such a path. The third configuration (to the right) differs from the middle one in only one transposition. It can therefore not be obtained by a local algorithm. Ergodicity breaking of the present type is very easily fixed, by simply considering a larger class of possible moves.
- The much more vicious ergodicity breaking appears when the algorithm is "formally correct", but is simply too slow. After a long simulation time, the algorithm may not have visited all the relevant corners of configuration space. The algorithm may be too slow by a constant factor N, or $\exp(N)$ .... Ergodicity breaking of this kind sometimes goes unnoticed for a while, because it may only show up clearly in particular variables, etc.. Very often, the system can be solidly installed in some local equilibrium, which does not however correspond to the thermodynamically stable state. This always invalidates the Monte Carlo simulation. There are many examples of this type of ergodicity breaking, e.g. in the study of disordered systems. Notice that the abstract proof that no "trivial" accident happens does not protect you from a "vicious" one.

For *orientation* (and knowing that I may add to the confusion) I should want to warn the reader lest he should think that in a benign simulation *all* the configurations have some reasonable chance to be visited. This is not at all the case, even for small systems. Using the very fast algorithm for the Ising model which I presented in the last chapter, you may generate for yourself energy histograms of, say, the 20×20 Ising model at two slightly different temperatures (above the Curie point). Even for very small differences in temperature, there are many configurations which have practically no chance ever to be generated at a given temperature, but which are commonly encountered at another temperature. This is of course the legacy of the Metropolis method where we sample configurations according to their statistical weight $\pi(x) \sim \exp[-\beta E(x)]$. This socalled "Importance sampling" is the saviour (and the danger) of the Monte Carlo method - but is related only to the equilibrium properties of the process. Ergodicity breaking - a sudden slowdown of the simulation - may be completely unrelated to changes in the equilibrium properties of the system. There are a few models in which this problem is discussed. In my opinion, none are as accessible as [8] which deals with a modified Ising model, that undergoes a purely dynamic roughening transition.

We maintain that the absolute probability $p(a \to \ldots \to b) > 0$ can very rarely be evaluated explicitly, and that the formal mathematical analysis is useful only to detect the "trivial" kind of ergodicity violation. Very often, careful data analysis and much physical insight are needed to assure us of the practical correctness of the algorithm.

Dynamics One is of course very interested in the numerical study of the phenomena associated with very slow dynamics, such as relaxation close to phase transitions, as glasses, spin glasses, etc.. We have just seen that these systems are *a priori* difficult to study with Monte Carlo methods, since the stationary distribution is never reached during the simulation.

It is characteristic of the way things go in Physics that - nonwithstanding these difficulties - there is often no better method to study these systems than to do a ... dynamical Monte Carlo simulation! In *dynamical Monte Carlo*, one is of course bound to a *specific* Monte Carlo algorithm, which serves as a *model* for the temporal evolution of the system from one local equilibrium state to another. In these cases one knows by construction that the Monte Carlo algorithm will drive the system towards the equilibrium, but very often after a time far too large to be relevant for experiment and simulation. So one is interested in studying the relaxation of the system from a given initial configuration.

The conceptual difference between equilibrium Monte Carlo (which was treated in the last chapter) and dynamical Monte Carlo methods cannot be overemphasized. In the former we have an essentially unrestricted choice of algorithm (expressed by the a priori probability, which was discussed at length), since one is interested exclusively in generating independent configurations x distributed according to $\pi(x)$. In thermodynamic Monte Carlo, the temporal correlations are just a nuisance. As we turn to dynamical calculations, these correlations become the main object of our curiosity. In dynamical simulations, the a priori probability is of course fixed. Also, if the Monte Carlo algorithm is ergodic both in principle and in practice, the static results will naturally be independent of the algorithm.

You may ask whether there is any direct relationship between Metropolis dynamics (which serves as our *model*), and the true physical time of the experiment, which would be obtained by a numerical integration of the equations of motion, as are done for example in molecular dynamics. There has been a lot of discussion of this point and many simulations have been devoted to an elucidation of this question for, say, the hard sphere liquid. All these studies have confirmed our intuition (as long as we stay with purely local Monte Carlo rules): the difference between the two approaches corresponds to a renormalization of time, as soon as we leave a ballistic regime (times large compared to the mean-free time). Monte Carlo dynamics are very often simpler to study.

In equilibrium Monte Carlo, theory does not stop with the naive Metropolis algorithm. Similarly, in a dynamical simulation there is room for much algorithmic subtlety. In fact, even though our *model* for the dynamics is fixed, we are not forced to implement the Metropolis rejections blindly on the computer. Again, it is the rate of rejections which serves as an important indicator that something more interesting than the naive Metropolis algorithm may be tried. The keyword here is *faster than the clock algorithms* which are surprisingly little appreciated, even though they often help to give birth to simulations of unprecedented speed.

2.2 Playing dice

As a simple example which is easy to remember, consider the system shown in Fig. 15: a single spin, which can be $S = \pm 1$ in a magnetic field $\mathcal{H}$, at finite

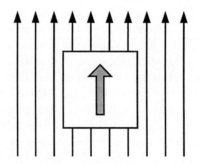

Fig. 15. Here, we look at the dynamics of a single spin in a magnetic field. We will soon find that it is related ...

temperature. The energy of each of the two configurations is

$$E = -\mathcal{H} S. \tag{27}$$

We consider the Metropolis algorithm of (10) to model the dynamical evolution of the system and introduce an explicit time step $\Delta\tau = 1$.

$$p(S \to -S, \Delta\tau) = \Delta\tau \begin{cases} 1 & \text{(if } E(-S) < E(S)) \\ \exp[-\beta(E(-S) - E(S))] & \text{(otherwise)} \end{cases}. \tag{28}$$

To be completely explicit, we write down again how the spin state for the next time step is evaluated in the Metropolis rejection procedure: at each time step

$t = 1, 2, \ldots$ we compute a random number $0 < ran < 1$ and compare it to the probability p from (28):

$$S_{t+1} = \begin{cases} -S_t & \text{if } p(S_t \to -S_t) > \text{ran} \\ S_t & \text{otherwise} \end{cases} \qquad (29)$$

This rule insures that asymptotically the two configurations are chosen according to the Boltzmann distribution. Notice that whenever we are in the excited "−" state of the single spin model, our probability to fall back on the next time step is 1, which means that the system will flip back to the "+" state on the next move. Therefore, the simulation will look like this:

$$\ldots + + + + \boxed{- +} + \boxed{- +} + + \boxed{- +} + + + + \boxed{- +} + + \boxed{- +} \ldots \qquad (30)$$

As the temperature is lowered, the probability to be in the excited state will gradually diminish, and you will spend more and more time computing random numbers in (29), but rejecting the move from $S = +1$ to $S = -1$.

At the temperature $\beta = \log(6)/2$, the probability to flip the spin is exactly $1/6$, and the physical problem is then identical to the game of the little boy depicted in Fig. 16. He is playing with a special die, with 5 empty faces (cor-

Fig. 16. ...to the game of this little boy. There is a simpler way to simulate the dynamics of a single spin than to throw a die at each time step. The solution is indicated in the lower part: the shaded area corresponds to the probability of obtaining a "flip" on the third throw.

responding to the rejections $S_{t+1} = S_t = 1$) and one face with the inscription "flip" ($S_{t+1} = -S_t$, $S_{t+2} = -S_{t+1}$). The boy will roll the die over and over again, but of course most of the time the result of the game is negative. As the game goes on, he will expend energy, get tired, etc. etc., mostly for nothing. Only very rarely does he encounter the magic face which says "flip". If you play this game in real life or on the computer, you will soon get utterly tired of all these calculations which result in a rejection, and you might start thinking that there must be a more economical way to arrive at the same result. In fact, you can predict analytically what the distribution of time intervals between "flips" will be. For the little boy, at any given time, there is a probability of 5/6 that one roll of the die will result in a rejection, and a probability of $(5/6)^2$ that two rolls result in successive rejections, etc.. The numbers $(5/6)^l$ are shown in the lower part of Fig. 16. You can easily convince yourself that the shaded area in the figure corresponds to the probability $(5/6)^2 - (5/6)^3$ of having a flip at exactly the third roll. So, to see how many times you have to roll to get a "flip", you simply draw a random number ran $0 < ran < 1$, and check into which box it falls. The box index l is easily calculated:

$$(5/6)^{l+1} < ran < (5/6)^l \Rightarrow l = \text{Int}\left\{\frac{\log ran}{log(5/6)}\right\}. \tag{31}$$

What this means is that there is a very simple program "dice.f", which you can obtain from my Web site, and which has the following characteristics:

- the program has no rejections. For each random number drawn, the program computes a true event: the waiting time for the next flip.
- The program is thus "faster than the clock" in that it is able to predict the state of the boy's log-table at simulation time t with roughly t/6 operations.
- the output of the accelerated program is completely indistinguishable from the output of the naive Metropolis algorithm.

You can see that to determine the temporal evolution of the die-playing game you don't have to do a proper "simulation", you can use a far better algorithm.

It is of course possible to generalize the little example from a probability $\lambda = 1/6$ to a general value of λ, and from a single spin to a general statistical mechanics problem with discrete variables. ...I hope you will remember the trick next time you are confronted with a simulation and the terminal output indicates one rejection after another. So you will remember that the proper rejection method, (29), is just one possible implementation of the Metropolis algorithm (28). The method has been used to study different versions of the Ising model, disordered spins, kinetic growth and many other phenomena. At low temperatures, when the rejection probabilities increase, the benefits of this method can be enormous.

So, you will ask why you have never heard of this trick before. One of the reason for the relative obscurity of the method can already be seen on the one-spin example: in fact you can see that the whole method does not actually use the factor 1/6, which is the probability to do something, but $5/6 = 1 - 1/6$, which is the probability to do nothing. In a general spin model, you can flip each of the N spins. As input to your algorithm computing waiting times, you again need the probability "to do nothing", which is

$$1 - \lambda = 1 - \sum_{1}^{N} [\text{probability to flip spin } i]. \tag{32}$$

If these probabilities all have to be computed anew for each new motion, the move becomes quite expensive (of order N). A straightforward implementation therefore has a good chance to be too onerous to be extremely useful. In practice, however, you may encounter enormous simplifications in evaluating λ for two reasons:

- You may be able to use symmetries to compute all the probabilities. Since the possible probabilities to flip a spin fall into a finite number n of *classes*. The first paper on accelerated dynamical Monte Carlo algorithms [9] has coined the name "n-fold way" for this strategy.
- You may simply be able to *look up* the probabilities, instead of computing them, if you are not visiting the present configuration for the first time and you have taken notes [12].

2.3 Accelerated algorithms for discrete systems

We now discuss the main characteristics of the method, as applied to a general spin model with configurations S_i made up of N spins $S_i = (S_{i,1}, \ldots, S_{i,N})$. We denote the configuration obtained from S_i by flipping the spin m by $S_i^{[m]}$. The system remains in configuration S_i during a time τ_i, so that the time evolution can be written as

$$S_1(\tau_1) \to S_2(\tau_2) \ldots \to S_i(\tau_i) \to S_{i+1}(\tau_{i+1}) \ldots etc.. \tag{33}$$

The probability of flipping spin m is given by

$$p(S_i \to S_i^{[m]}, \Delta\tau) = \frac{\Delta\tau}{N} \begin{cases} 1 & (\text{if } E(S_i^{[m]}) < E(S_i)) \\ \exp[-\beta(\Delta E)] & (\text{otherwise}) \end{cases}, \tag{34}$$

where the $1/N$ factor expresses the $1/N$ probability to select spin m. After computing $\lambda = \sum_i p(S_i \to S_i^{[m]})$, we obtain the waiting time as in (31), which gives the exact result for finite values of $\Delta\tau$. Of course, in the limit $\Delta\tau \to 0$ (31) simplifies, and we can sample the time to stay in S_i directly from an exponential distribution $p(\tau_i) = \lambda \exp(-\lambda \tau_i)$ (cf. [1] chap. 7.2 for how to generate exponentially distributed random numbers).

If we have then found that after a time τ we are going to move on from our configuration S, *where* are we going to move to? The answer to this question can be easily understood by looking at Fig. 17: there we see the "pile" of all the probabilities which were computed. If the waiting time τ_i is obtained from λ_i, we choose the index $[m]$ of the flipped spin with a probability $p(S \to S^{[m]})$.

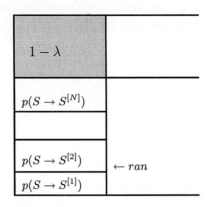

Fig. 17. "Pile" of probabilities, which allows us to understand how we decide where to go. Accelerated algorithms make sense if, generally, $\lambda \ll 1$.

To do this, you need all the elements to produce the box in Fig. 17 i. e. the probabilities:

$$
\begin{aligned}
& p(S \to S^{[1]}) + p(S \to S^{[2]}) + \ldots p(S \to S^{[N]}) = \lambda \\
& \quad \vdots \\
& p(S \to S^{[1]}) + p(S \to S^{[2]}) \\
& p(S \to S^{[1]})
\end{aligned}
\tag{35}
$$

In addition, one needs a second random number ($0 < ran < \lambda$) to choose one of the boxes (cf the problem in Fig. 16). The best general algorithm to actually compute m is of course not "visual inspection of a postscript figure", but what is called "search of an ordered table". This you will find explained in any book on basic algorithms (cf, for example [1], chap. 3.2). Locating the correct box only takes order $\log(N)$ operations. The drawback of the computation therefore is that any move costs order N operations, since in a sense we have to go through all the possibilities of doing something before knowing our probability "to do nothing". This is the price to be payed for eliminating rejections altogether.

2.4 Implementing accelerated algorithms

Once you have understood the basic idea of accelerated algorithms, you may wonder how these methods are implemented, and whether it's worth the trouble. In cases in which the local energy is the same for any lattice site, you will find that the probabilities can only take on n different values. In the case of the 2-dimensional Ising model, there are 10 classes that the spin can belong to, ranging from up-spin with 4 neighboring up-spins, up-spin with 3 neighboring up-spins, to down-spin with 4 neighboring down-spins. Knowing the repartition into different classes at the initial time, you can see that flipping the spin changes the classes of 5 spins, and can be seen as a change of the number of particles belonging to the different classes. Using somewhat more intricate bookkeeping, we can therefore compute the value of λ in a constant number of operations, and the cost of making a move is reduced to order $O(1)$. You can see that the accelerated algorithm really has solved the problem of small acceptance probabilities which haunt so many simulations at low temperatures (for practical details, see refs [9], [10]).

...If you program the method, the impression of bliss may well turn into frustration, for we have overlooked an important point: the system's dynamics, while without rejections, may still be futile. Consider for the sake of concreteness an energy landscape such as in Fig. 18, where any configuration is connected to

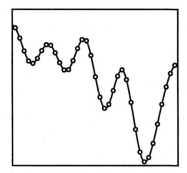

Fig. 18. Rugged energy "landscape" which poses problems in a dynamical Monte Carlo simulation at low temperatures. The system will be stuck in one of the local minima, and the dynamics will be futile, i. e. very repetitive.

two other configurations. Imagine the system at one of the local minima at some initial time. At the next time step, it will move to one of the neighboring sites, but it will almost certainly fall back right afterwards. At low temperatures, the system will take a very long time (and, more importantly, a very large number of operations) to hop over a potential barrier. In these cases, the dynamics is extremely repetitive, futile. If such behavior is forseeable, it is of course wasteful to recompute the "'pile of probabilities", (35), and even to embark on the involved book-keeping tricks of the n-fold way algorithm. In these cases, it is much more economical to save much of the information about the probabilities, and to look them up, when next needed. An archive can be set up in such a way that, upon making the move $S_i \to S_i^{[m]}$, we can quickly decide whether we have seen the new configuration before, and we can immediately look up the "pile of probabilities". This leads to extremely fast algorithms (for practical details, see [11], [12]). Systems in which this approach can be used are, amongst others: flux lines in a disordered superconducter, the NNN Ising model [8] alluded to earlier, disordered systems with a so-called single-step replica symmetry breaking transition, and in general systems with steep local minima. For these systems it is even possible to produce second-generation algorithms, which not only accept a move at every time step, but even move to a configuration which the system has never seen before. One such algorithm was proposed in [11].

There are endless possibilities to improve the dynamical algorithms for some systems. Of course, the more powerful a method, the less frequently it can be applied. It should be remembered that the above method is only of use when the probability to do nothing is very large, and/or if the dynamics is very repetitive (small local minimum of energy function in Fig. 18). A large class of systems for which none of these conditions hold are spin glasses with a so-called continuous replica symmetry breaking transition, such as the Sherrington-Kirkpatrick model. In these cases, the ergodicity breaking takes place very "gracefully": there are very many configurations accessible for any given initial condition. In this case, there seems to be very little room for improvements of the basic method.

2.5 Random sequential adsorption

I do not want to leave the reader with the impression that accelerated algorithms are restricted to spin models only. In fact, intricate rapid methods can be thought up for most dynamical simulations with a large numbers of rejections. These rejections simply indicate that the time to "do something" may be much larger than the simulation time step.

The following example, random sequential adsorption, was already mentioned in the previous section. Imagine a large two-dimensional square on which you deposit one coin each second - but attention: we only put the coin down if it does not overlap with any of the coins deposited before. The light gray coin in Fig. 19 will immediately have to be taken away. Random sequential adsorption is a particularly simple dynamical system because of its irreversibility. We are interested in two questions:

- The game will stop when it is not possible to deposit any more coins. What is the time after which this "jamming" state is reached and what are its properties?
- We would also like to know the mean density as a function of time for a large number of different realizations.

I will give a rather complete discussion of this problem here in six steps, displaying the panoply of refinements which can be unfolded as we come to understand the program. There is a large research literature on the subject, but you will see that you can find an optimal algorithm yourself by simply applying what we have learnt about the problem of the rolling die.

Naive algorithm You can write a program to simulate this in a matter of minutes. You need a table which contains the (x, y) positions of the $N(t)$ coins

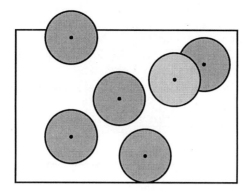

Fig. 19. Random sequential adsorption: the light grey coin - which has just been deposited - has to be taken away again.

deposited previously and a random number generator which will give the values of x, y. If you run the program you will see that even for modest sizes of the box, it will take quite a long time before it will stop. You may say "Well, the program is slow simply because the physical process of random absorption is slow, there is nothing I can do ...". If that is your reaction, you may gain something from reading on. You will find that there is a whole cascade of improvements that can be incorporated into the program. These improvements concern not only the implementation details but also the deposition process itself, which can be simulated faster than in the experiment - especially in the final stages of the deposition. Again, there is a *faster than the clock* algorithm, which deposits (almost) one particle per unit time. These methods had been explored very little in the past.

Underlying lattice The first thing you will notice is that the program spends a lot of time computing distances between the proposed point (x, y) and the coins which have already been deposited. It is evident that you will gain much time by performng only a local search. This is done with a grid, as shown in Fig. 20 and by computing the table of all the particles contained in each of the little squares. When you try to deposit the particle at (x, y), you first compute the square which houses the point (x, y) and then compute the overlaps with particles contained in neighboring squares. Some book-keeping is necessarily involved, and varies with the size of the squares adopted. There has been a lot of discussion about how big the little squares have to be taken, and there is no clear-cut answer. Some people prefer a box of size approximately $\sqrt{2}$ times the radius of the spheres. In this case you can be sure that at most 1 sphere per square is present, but the number of boxes which need to be scrutinized is quite large. Others have adopted larger boxes which have the advantage that only the contents of 9 boxes have to

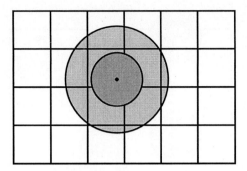

Fig. 20. One of the basic techniques to orient oneself is to introduce a grid. The exlusion area of the coin is also shown.

be checked. In any case, one gains an important factor N when compared with the naive implementation.

Stopping criterium Since we have said that we want to play the game up to the bitter end, you may want to find out whether there is any possibility of depositing one more coin at all. The best thing to do is to write a program which will tell you whether the square can host one more point.

Excluding squares Using the idea of the exclusion disk, you will be able to compute which parts of the whole field can take one more coin. These areas are shaded dark and we shall call them "stars" for obvious reasons. You can

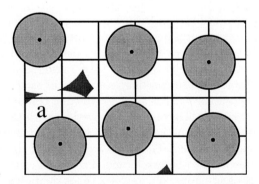

Fig. 21. Even though there is much empty space, we can only deposit further coins in three "stars" (extending to 4 of the 24 squares making up the field).

see that in the situation depicted in the figure only 4 of the 24 grid-squares can take on a new point. Before starting to compute distances from coins in adjacent squares, it may be a good idea to check at all whether it is possible to deposit a box on the point. You will quickly realize that we are confronted with exactly the same problem as the boy in Fig. 16, and that the time for the next hit of one of the useful squares can be sampled from (31). So, you can write a faster (intermediate) program, by determining which of the boxes can still take a coin. This probability then gives the probability "to do nothing", which is used in (31) to sample the time after which a new deposition is attempted. Are you tempted to write such a program? You simply need a subroutine which determines whether there still is free space in a given square. With such a subroutine you are able to exclude squares from the consideration. The ratio of excluded squares to the total number of squares then gives the probability $1 - \lambda$ to do nothing, which is what you need for the faster-than-the-clock algorithm of the last section.

The ultimate algorithm This trick will allow you to make the program run faster by a factor N, where N is the number of little squares. Unfortunately, you will quickly find that the program still has a very large rejection probability... just look at the square denoted by an a) in Fig. 16: roughly 2% of the square's surface can accept a new coin. So you will attempt many depositions in vain before being able to do something reasonable. You can try pushing these ideas further by using smaller and smaller squares. This was implemented in the literature in [13]. What one really wants to do, however, is to exploit the exact analogy between the area of the stars, and the probabilities in (35). If we know the locations and the area of the stars, we are able to implement one of the rejection-free algorithms. Computing the area of a "star" is a simple trigonometric exercise. Having such a subroutine at our disposal allows us to envision the *ultimate program* for random sequential adsorption.

- Initially, you do the naive algorithm for a while.
- Then you do a first cutting up into stars. The proportion of the field not covered by stars correspons to the factor λ, and you will sample the star for the next deposition exactly as in (35). Then you randomly deposit (x, y) into the star chosen, and update the book-keeping.

How to sample a random point in a star So, finally, we have reduced the problem of adsorption to the problem of random deposition of a coin in a star. How do you do that[8]? As for me, I have given up looking for a rejection-free method to solve this problem...but do you perhaps know how to do this?

Literature, extensions Of course, the example of random adsorption was only given to stimulate you to think about better algorithms for *your* current Monte

---

[8] Question not treated in "Le Petit Prince" by A. de St Exupery

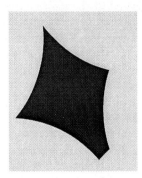

Fig. 22. The ultimate algorithm for random sequential adsorption needs a program to calculate the area of a star (straightforward), and a method to sample a random point within it. Can you sample random points within the star without rejections?

Carlo problem, and how it may be possible in your own research problem to get away from the blind use of algorithms. If you want to know more about random deposition, consult the research literature which is enormous, and refer to the algorithm presented in [13]. Notice that in our algorithm it was very important for the spheres to be monodisperse, i.e. for them all to have the same diameter. What could we do if this was not the case? Are there any accelerated algorithms for spheres with some distribution of diameters (from d_{min} to d_{max} (easy), and what would be an optimal algorithm for the deposition of objects with additional degrees of freedom? The problem is of some interest in the case of ellipses. Evidently, from a numerical point of view, you will end up with a three-dimensional "star" in which you have to sample (x, y, θ), where θ gives the orientation of the ellipse to be deposited. You may feel tempted to think about such a simulation. Remember that it is not important to compute the 3d star exactly, just as, in the last chapter, it was not essential to make the a priori probability $\mathcal{A}(x)$ exactly $\pi(x)$.

References

[1] Press, W. H., Teukolsky, S. A., Vetterling, W. T., Flannery, B. P., *Numerical Recipes*, 2nd edition, Cambridge University Press (1992).
[2] Lee, J., Strandburg K. J., *Phys. Rev.* **B 46** 11190 (1992).
[3] Pollock, E: L., Ceperley, D. M., *Phys. Rev.* **B 30**, 2555 (1984), **B 36** 8343 (1987); Ceperley, D. M, *Rev. Mod. Phys.* **67**, 1601 (1995).
[4] Caracciolo, S., Pelissetto, A., Sokal, A. D., *Phys. Rev. Lett* **72** 179 (1994).
[5] Swendsen, R. H., Wang, J.-S., *Phys. Rev. Lett.***63**, 86 (1987).
[6] Wolff, U., *Phys. Rev. Lett.* **62**, 361 (1989).
[7] Ferdinand A. E., Fisher, M. E., *Phys. Rev.* **185** 185 (1969).

[8] Shore, J. D., Holzer, M., Sethna, J. P., *Phys Rev.* **B 46** 11376 (1992).
[9] Bortz, A. B., Kalos, M. H., Lebowitz, J. L., *J. Comput. Phys.* **17**, 10 (1975) *cf* also : Binder, K., in *Monte Carlo Methods in Statistical Physics*, edited by K. Binder, 2nd ed. (Springer Verlag, Berlin, 1986, sect 1.3.1).
[10] Novotny, A. M., *Computers in Physics* **9** 46 (1995).
[11] Krauth, W., Pluchery, O., *J. Phys. A: Math Gen* **27**, L715 (1994).
[12] Krauth, W., Mézard, M., *Z. Phys.* **B 97** 127 (1995).
[13] Wang, J.-S., *Int. J. Mod. Phys.* **C 5**, 707 (1994).

Cluster Algorithms

Ferenc Niedermayer

Institute for Theoretical Physics, University of Bern, Sidlerstrasse 5, CH-3012 Bern[*]

Abstract. Cluster algorithms for classical and quantum spin systems are discussed.

1 Introduction

The simulation of statistical systems close to their critical points is usually a very difficult problem, because the dynamic evolution of the system in Monte Carlo (or computer-) time slows down considerably. This phenomenon is called critical slowing down (CSD). It also occurs in real systems (in real time) and is an interesting subject to study. In Monte Carlo (MC) simulations, however, where one is interested in quantities averaged over the equilibrium probability distribution, CSD can lead to a tremendous waste of computer time. The evolution of a system is characterized by the autocorrelation time – this defines the rate at which the system loses memory of a previous state, or in other words, the number of MC steps needed to generate a statistically independent new configuration.

Let us consider an observable A (a function of spins if we consider a spin system) and denote by A_t its value at a given MC time t. We define the *autocorrelation time* through the average over the equilibrium distribution as

$$C_{AA}(t) = \langle A_s A_{s+t}\rangle - \langle A\rangle^2 \qquad (1)$$

For large t this decays exponentially, $C_{AA}(t) \propto \exp(-t/\tau_{exp,A})$ where $\tau_{exp,A}$, the *exponential autocorrelation time* corresponds to the slowest mode in the MC dynamics. It is also useful to define a slightly different quantity, the *integrated autocorrelation time*

$$\tau_{int,A} = \frac{1}{2}\sum_{t=-\infty}^{+\infty} C_{AA}(t)/C_{AA}(0) = \frac{1}{2} + \sum_{t=1}^{+\infty} C_{AA}(t)/C_{AA}(0) \qquad (2)$$

This is the quantity which is relevant for the statistical error in a MC simulation. One usually estimates $\langle A\rangle$ from the average of n subsequent measurements, i.e. through the quantity

$$\bar{A} = \frac{1}{n}\sum_{t=1}^{n} A_t \qquad (3)$$

Its statistical error, $\delta\bar{A}$ for $n \gg \tau_{int,A}$ is given by

$$\delta\bar{A} = \sqrt{2\tau_{int,A}} \cdot \delta\bar{A}_{naive} \qquad (4)$$

---

[*] On leave from the Institute of Theoretical Physics, Eötvös University, Budapest

where
$$\delta \bar{A}_{naive} = \sqrt{\frac{1}{n}C_{AA}(0)} \qquad (5)$$

Equation (4) means that in order to achieve a given accuracy one has to spend computer time $\propto \tau_{int}$. Unlike the equilibrium averages $\langle A \rangle$ or $C_{AA}(0)$, the autocorrelation time depends on the actual MC algorithm. In general, near a critical point (i.e. for $\xi \gg 1$) the autocorrelation time diverges as

$$\tau \approx c\xi^z \qquad (6)$$

where z is the *dynamic critical exponent*.[1]

For a local update, such as the standard Metropolis algorithm, one has $z \approx 2$. Some 2d systems reveal their critical properties only at quite large correlation length ξ. In two dimensions the memory allows consideration of systems with linear size $L \approx 1000$ and a correlation length $\xi \approx 100-200$. For a local algorithm, however, (6) predicts $\tau \sim 10000$, which makes simulation with a local Metropolis algorithm hopeless.

The $\tau \propto \xi^2$ behaviour reminds one of the random walk — a change at some site will propagate to a distance of $\sim \xi$ due to random local updating steps in a time proportional to ξ^2.

There are several ways to change the MC dynamics in order to decrease the value of z. One way is to perform a deterministic local update instead of a random one, staying on the energy surface E=const, i.e. making a micro-canonical step. This *over-relaxation algorithm* [2] can lower the value of the dynamic critical exponent down to $z \approx 1$.

The other class of MC dynamics updates a collective mode instead of a single local variable. In some optimal cases one can reach $z \approx 0$, i.e. completely eliminate (at least from a practical point of view) the problem of CSD. Two important methods of this type are

- Multi-grid algorithms
- Cluster algorithms

In the multi-grid algorithms one introduces in addition to the original (fine grid) lattice a sequence of lattices with lattice spacings $a' = 2a$, $a'' = 4a$, ... and updates the corresponding block of spins. i.e. regions of given sizes and shapes. For the $O(N)$ vector model the multi-grid algorithm reduces the dynamic critical exponent to $z \approx 0.5 - 0.7$ [3]. As we shall see, the cluster algorithms – the topics of these notes – work better for this system. One should keep in mind, however, that for other systems the presently available cluster algorithms are not efficient, while the multi-grid algorithms can still be used. In these lecture notes I discuss the cluster algorithms for classical and quantum spin systems.

---

[1] For more details and for references on statistical time–series analysis see e.g. [1].

2 Cluster Algorithm for Classical Spin Systems

The cluster MC method has been suggested by Swendsen and Wang [4] for the Ising model, based on an earlier observation that the partition function of the system can be written as a sum over cluster distributions [5]. In Ref. [6] the algorithm has been generalized to spin systems, using more general considerations. Here I follow this derivation since it will be straightforward to apply it to general $O(N)$ lattice actions.

Consider a general $O(N)$ vector model where the energy of the configuration $S = \{\mathbf{S}_n\}$ is given by[2]

$$E(S) = \sum_l E_l(S) \ . \tag{7}$$

Here l denotes a given link, a pair of sites $l = (n_1, n_2)$ and the interaction term $E_l(S)$ depends only on the corresponding spins, $E_l(S) = E_l(\mathbf{S}_{n_1}, \mathbf{S}_{n_2})$. At this point let us extend the class of actions considered in Ref. [6] and consider the case of 3-spin (4-spin, ...) interaction terms, and let l be a (hyper)link $l = (n_1, n_2, n_3)$. The corresponding interaction term is $E_l(S) = E_l(\mathbf{S}_{n_1}, \mathbf{S}_{n_2}, \mathbf{S}_{n_3})$ (and similarly for 4-spin, ... interactions). We shall also assume that the interaction has a global symmetry,

$$E_l(g\mathbf{S}_1, g\mathbf{S}_2, \ldots) = E_l(\mathbf{S}_1, \mathbf{S}_2, \ldots) \ , \tag{8}$$

for $g \in G$ where G is the corresponding symmetry group. To construct the clusters we connect the sites belonging to some link l by a 'bond' with some probability $p_l(S)$ depending on the spins associated with the link. We assume it to be globally invariant as well:

$$p_l(g\mathbf{S}_1, g\mathbf{S}_2, \ldots) = p_l(\mathbf{S}_1, \mathbf{S}_2, \ldots) \ . \tag{9}$$

The probability to produce a given configuration of bonds B is given by

$$w(B|S) = \prod_{l \in B} p_l(S) \prod_{l \notin B} (1 - p_l(S)) \ . \tag{10}$$

The next step is to build the clusters. The cluster is the set of sites which can be visited from each other going through bonds. We propose now the following change in the configuration: spins within a given cluster are transformed globally, by some $g_i \in G$, $i = 1, \ldots, N_C$, where N_C is the actual number of clusters defined by the bond configuration. (Note that different bond configurations can result in the same cluster configuration. In fact, only the cluster configuration matters here.) The equation of detailed balance (assuming the same a priori probability for g_i^{-1} as for g_i) is given by

$$\mathrm{e}^{-E(S)} w(B|S) w(S \to S') = \mathrm{e}^{-E(S')} w(B|S') w(S' \to S) \ . \tag{11}$$

---

[2] For simplicity, we include the factor $\beta = 1/kT$ in the definition of E, i.e. the Boltzmann factor is $\exp(-E)$

Here $w(S \to S')$ and $w(S' \to S)$ are the corresponding acceptance probabilities used to correct the equation (see Ref. [7]). Because of the global symmetry, (8,9), the contributions coming from a link l with all its sites belonging to the same cluster cancel, we have

$$\prod_{l \notin B} e^{-E_l(S)}(1 - p_l(S))w(S \to S') = \prod_{l \notin B} e^{-E_l(S')}(1 - p_l(S'))w(S' \to S) \ . \quad (12)$$

Let us introduce a convenient parametrization for $p_l(S)$:

$$p_l(S) = 1 - e^{E_l(S) - Q_l(S)} \ . \quad (13)$$

The fact that $p_l(S)$ is a probability requires

$$Q_l(S) \geq E_l(S) \ . \quad (14)$$

Equation (12) gives then

$$\prod_{l \notin B} e^{-Q_l(S)} w(S \to S') = \prod_{l \notin B} e^{-Q_l(S')} w(S' \to S) \ . \quad (15)$$

Here in fact only those links contribute which would connect different clusters, i.e. are on the 'surface' of clusters. We have succeeded in mapping our original spin system onto a system of N_C 'sites' (representing the clusters $1, \ldots, N_C$) with the dynamic variables $g_i, i = 1, \ldots, N_C$. The interaction between the clusters is given by $Q_l(S)$ where l is a (hyper)link connecting two or more clusters. (Note that by taking $Q_l(S) = E_l(S)$, i.e. $p_l(S) = 0$, each lattice site becomes a cluster by itself and we recover the local MC method.)

Let us specify the bond probabilities, i.e. $Q_l(S)$. We restrict the proposed transformations g to a subgroup $H \subset G$ to be specified later, and let $Q_l(S)$ be the maximal energy $E_l(S')$ where the elements $g \in H$ are applied independently to all possible sites belonging to l,

$$Q_l(\mathbf{S}_1, \mathbf{S}_2, \ldots) = \max_{g_1, g_2, \ldots \in H} E(g_1 \mathbf{S}_1, g_2 \mathbf{S}_2, \ldots) \ . \quad (16)$$

Clearly, we have $Q_l(g_1 \mathbf{S}_1, g_2 \mathbf{S}_2, \ldots) = Q_l(\mathbf{S}_1, \mathbf{S}_2, \ldots)$ for any $g_1, g_2, \ldots \in H$. For this choice of the bond probabilities the remaining factors $l \notin B$ in (15) cancel and one obtains

$$w(S \to S') = w(S' \to S) \ . \quad (17)$$

This is a remarkable fact: the accept-reject step usually needed to correct the equation of detailed balance is unnecessary, the clusters can be updated independently. This is possible because the clusters are built dynamically, i.e. they are sensitive to the interaction and to the actual configuration. With a fixed a priori given shape of clusters one cannot achieve this.

The construction is still quite general, but there is a hidden subtle point here. If $Q_l - E_l$ is too large then the bond probability becomes too large as well. As a consequence, the largest cluster will usually occupy almost the whole lattice, except for a few small isolated clusters. Although the algorithm is still correct, it is also useless in this case: applying a global transformation to the whole lattice does not change the relative orientation of the spins, so one would effectively update a few small clusters with the computation being done on the whole lattice. Before proceeding with the general case of the O(N) spin model, let us consider the Ising model as a special case.

$$E = -J \sum_{n,\mu} S_n S_{n+\hat{\mu}} , \tag{18}$$

where $J > 0$ and $S_n = \pm 1$. The global symmetry transformation in this case is $S_n \to g S_n$, $g = \pm 1$. Equation (16) gives

$$Q_l(S) = J , \tag{19}$$

and consequently

$$p_l(S_1, S_2) = \begin{cases} 1 - e^{-2J} & \text{for parallel spins, } S_1 S_2 = +1 , \\ 0 & \text{for antiparallel spins, } S_1 S_2 = -1 . \end{cases} \tag{20}$$

This is the bond probability for the Swendsen–Wang algorithm. An important observation is that the clusters cannot grow too large: since the bond probability is zero between antiparallel spins, all the spins in a cluster will have the same sign – the clusters cannot grow larger than the region of same-sign spins. In the unbroken phase the latter does not exceed half of the total number of sites, hence the main danger of cluster algorithms – that one ends up with a single cluster occupying nearly the whole volume – is avoided in this case.

Since the clusters do not interact (c.f. (17)) one can choose any of the 2^{N_C} sign assignments with equal probability. One can average over these 2^{N_C} possibilities without actually doing the updates – by introducing the *improved estimators*. For the correlation function one has

$$\langle S_x S_y \rangle = \sum_\mathcal{C} p(\mathcal{C}) \langle S_x S_y \rangle_\mathcal{C} , \tag{21}$$

where $\mathcal{C}$ is a given cluster distribution appearing with probability $p(\mathcal{C})$ and $\langle S_x S_y \rangle_\mathcal{C}$ is the average over the 2^{N_C} possibilities for $\mathcal{C}$. The latter is trivial,

$$\langle S_x S_y \rangle_\mathcal{C} = \delta_{xy}(\mathcal{C}) , \tag{22}$$

where $\delta_{xy}(\mathcal{C})$ is 1 if the two sites belong to the same cluster, and 0 otherwise. Hence we have

$$\langle S_x S_y \rangle = \langle \delta_{xy}(\mathcal{C}) \rangle . \tag{23}$$

The improvement comes from the fact that at large distances $|x - y|$ the small value of correlator $\langle S_x S_y \rangle \ll 1$ is obtained by averaging over values $+1$ and -1 in the standard measurement while in the case of the improved estimator $\delta_{xy}(\mathcal{C})$

it comes from averaging over +1 and 0. In other words, the value of $C_{AA}(0)$ in (5) is much smaller for the improved estimator than for the original one since

$$\langle (S_x S_y)^2 \rangle = 1, \quad \langle (\delta_{xy}(\mathcal{C}))^2 \rangle = \langle \delta_{xy}(\mathcal{C}) \rangle \ll 1 \ . \tag{24}$$

As a consequence, one expects to gain an additional factor $1/\langle S_x S_y \rangle$ in computer time. This argument is, unfortunately, not complete. The point is that one usually measures the correlation function $\langle S_x S_y \rangle$ from an average over the whole lattice with $|x-y|$ fixed. When the site x runs over the lattice the quantity $S_x S_y$ fluctuates more independently than the improved estimator $\delta_{xy}(\mathcal{C})$ – the values of the latter are strongly correlated when large clusters are present.

A variant of this algorithm has been introduced by Wolff [8] – the single-cluster algorithm. He suggested building just one cluster starting from a random site, reflecting all spins in this cluster and starting the procedure again. The difference between this and the original Swendsen–Wang multi-cluster algorithm is that one enhances the probability of updating large clusters since a cluster of size $|C|$ is hit with the probability $|C|/V$, where V is the total number of lattice sites. Large clusters still evolve too slowly compared to small ones in the multi-cluster method – the single-cluster version corrects for this by updating the large clusters more often. (Obviously, one can try to vary the maximal number of clusters to be updated within a given MC run, and optimize in the distribution of this number.)

The improved estimator can be modified for the case of the single-cluster algorithm [8]. For example, the correlation function is given by

$$\frac{1}{V}\sum_{x=1}^{V}\langle S_x S_{x+r}\rangle = \left\langle \sum_{i=1}^{N_C} \frac{|C_i|}{V} \frac{1}{|C_i|} \sum_{x\in C_i} \delta_{x,x+r}(\mathcal{C}) \right\rangle_{\mathrm{mc}} = \left\langle \frac{1}{|C_1|} \sum_{x\in C_1} \delta_{x,x+r}(\mathcal{C}) \right\rangle_{\mathrm{sc}}, \tag{25}$$

where the subscripts mc, sc refer to multi-cluster and single-cluster algorithms, respectively. The susceptibility in the unbroken phase is given by

$$\chi = \frac{1}{V}\sum_{x,y}\langle S_x S_y\rangle = \left\langle \frac{1}{V}\sum_{i=1}^{N_C}\sum_{x,y\in C_i}\delta_{xy}(\mathcal{C})\right\rangle = \left\langle \sum_{i=1}^{N_C}\frac{|C_i|}{V}|C_i|\right\rangle = \langle |C|\rangle_{\mathrm{sc}} \ . \tag{26}$$

The average size of a randomly chosen cluster diverges together with the susceptibility when one approaches the critical point – the MC dynamics adjusts itself to the long range correlations present in the equilibrium situation.

In applying the previous considerations to the $O(N)$ spin model one has to choose the proper symmetry transformation H in (16). Choosing this to be the rotation group – the choice made in [6] – (16) leads to a bond probability which is too large, the largest cluster tends to occupy the whole lattice. It has been suggested there to make only small rotations and consider the case of interacting clusters as in (15). A much better solution suggested by Wolff [8] is to choose the subgroup of reflections with respect to some given random direction $\mathbf{r}$. (The derivation in Ref. [8] was based on embedding an Ising model into the $O(N)$

model and applying to it the Swendsen-Wang cluster algorithm. The advantage of the present approach shows up in applications to O(N) lattice actions of more general type.) The corresponding subgroup contains the identity and the reflection of the parallel component of the spin to the vector **r**, i.e.

$$g: \quad S_n^\parallel \to -S_n^\parallel, \quad \mathbf{S}_n^\perp \to \mathbf{S}_n^\perp. \tag{27}$$

For the standard O(N) action we have

$$\mathbf{S}_n \mathbf{S}_{n+\hat\mu} = S_n^\parallel S_{n+\hat\mu}^\parallel + \mathbf{S}_n^\perp \mathbf{S}_{n+\hat\mu}^\perp, \tag{28}$$

and only the first term is affected by the update. Essentially we obtained an embedded Ising model with space-dependent *ferromagnetic* couplings,

$$E = -\sum_{n,\mu} J_{n,\mu} \epsilon_n \epsilon_{n+\hat\mu}, \quad \text{with } J_{n,\mu} = J|S_n^\parallel S_{n+\hat\mu}^\parallel|, \quad \epsilon_n = \text{sign} S_n^\parallel. \tag{29}$$

The corresponding bond probabilities are given by $Q_l = J_l$. Again, because of the ferromagnetic nature of the couplings, the regions with $\epsilon_n > 0$ and $\epsilon_n < 0$ are not connected by bonds and the size of the clusters is bounded by the size of the corresponding regions. Consequently (at least in the unbroken phase) one is safe from having clusters with $|C| \sim V$. For the O(N) vector model both the single-cluster [10] and multi-cluster [11] method practically eliminate the CSD. For this model the cluster algorithm works even better then for the Ising model where some CSD is still observed.

Note that the present formulation can be easily applied to more general O(N) actions. Let us consider first the Symanzik (tree-level) improved O(N) action [12],

$$E(S) = -J \sum_{n,\mu} \left[\frac{4}{3} \mathbf{S}_n \mathbf{S}_{n+\hat\mu} - \frac{1}{12} \mathbf{S}_n \mathbf{S}_{n+2\hat\mu} \right]. \tag{30}$$

Here we have two types of links, $l = (n, n+\hat\mu)$ and $l = (n, n+2\hat\mu)$, with $Q_l = \frac{4}{3}|S_n^\parallel S_{n+\hat\mu}^\parallel|$ and $Q_l = \frac{1}{12}|S_n^\parallel S_{n+2\hat\mu}^\parallel|$, respectively. A potential danger here is related to the fact that the second term in (30) is anti-ferromagnetic – the bonds of the second type can connect sites with opposite sign of $S_n^\parallel$. Fortunately, the coefficient $1/12$ is sufficiently small, and the largest cluster does not grow too large [13].

Another example of practical importance is provided by the classically perfect lattice action (or the fixed point action of a renormalization group transformation) [15] which has a form

$$E(S) = \frac{1}{2} \sum_{n,r} \rho(r)(1 - \mathbf{S}_n \mathbf{S}_{n+r}) + \sum_{n_1..n_4} c(n_1..n_4)(1 - \mathbf{S}_{n_1}\mathbf{S}_{n_2})(1 - \mathbf{S}_{n_3}\mathbf{S}_{n_4}) + ... \tag{31}$$

This lattice action is used to minimize the lattice artifacts (the discretization errors). Although introducing the signs $\epsilon_n = \text{sign}(S_n^\parallel)$ does not turn this action into an Ising model (couplings between more than two spins are also present) the discussion presented here readily applies to this general action, and the procedure practically eliminates the CSD.

One can also introduce an external magnetic field. The simplest way to apply the cluster algorithm is to consider the external field as an extra spin which is coupled to all spins, and consider this interaction term on the same footing as all other terms in $E(S)$[14]. The cluster to which the external spin belongs does not need to be updated, otherwise the procedure is unchanged.

When modifying the algorithm, the condition of detailed balance has to be rechecked carefully. To illustrate this, let us consider the 3d O(3) model in the broken phase. If the random direction **r** points towards the magnetization **M**, the effective couplings $J|S_n^\| S_{n+\hat{\mu}}^\||$ become too large and consequently the size of the largest cluster too. (This is expected physically – flipping half of the spins along the total magnetization does not produce a typical configuration.) Could one modify the algorithm by restricting **r** to be orthogonal to **M**? It is easy to see that this is not possible. The component $\mathbf{M}^\perp$ (orthogonal to **r**) remains unchanged by the updates while $M^\|$ changes from zero to some nonzero value hence $|\mathbf{M}|$ always grows. In fact, we violated detailed balance by biasing **r** with the direction of the magnetization. (On the other hand, it is permissible to take **r** orthogonal to the direction of a given spin.)

One can also define improved estimators for the O(N) spin model. Since only the signs of $S_n^\|$ are updated one can use the relation

$$\langle \mathbf{S}_x \mathbf{S}_y \rangle = N \langle S_x^\| S_y^\| \rangle \;, \tag{32}$$

and define the improved estimator as

$$N|S_x^\| S_y^\|| \, \delta_{xy}(\mathcal{C}) \;. \tag{33}$$

Unfortunately, this procedure introduces an unwanted noise in the correlator when $\mathbf{S}_x$ and $\mathbf{S}_y$ are correlated significantly (i.e. for $|x - y| \lesssim \xi$). To illustrate this let us consider $\mathbf{S}_x \mathbf{S}_x$: this is, of course, exactly 1 when measured directly, it has no statistical error while the estimator $N S_x^\| S_x^\|$ fluctuates. The problem is cured easily [16]: one should only choose N random directions $\mathbf{r}_1, \mathbf{r}_2, \ldots, \mathbf{r}_N$ in a sequence forming an *orthogonal system* and apply the cluster algorithm using $\mathbf{r} = \mathbf{r}_i$ for $i = 1, \ldots, N$, otherwise everything remains unchanged, including the improved estimator. Unfortunately, for the single-cluster update this trick is not very convenient. To cure the problem with the improved measurement and keep the advantage that the single-cluster method updates more efficiently the large clusters one can do the following:

1. make n_s single-cluster steps, but do not measure anything,
2. make N consecutive multi-cluster steps with a random basis and perform a measurement after each step.

One can try to optimize the value of n_s to get a uniformly small autocorrelation time at all scales or even try to vary the maximal number of clusters $(1, 2, \ldots)$ to be built and updated. We shall refer to this as the 'hybrid' method. To illustrate the performance of the cluster method and of the improved estimators I made some runs on a 256^2 lattice at $\beta = 1.8$ which corresponds to a correlation length $\xi = 64.78(15)$ [10]. For the hybrid algorithm I took $n_s = 25 \approx V/\langle|C|\rangle$, and measured the correlation function $C(x) = \langle \mathbf{S}_0 \mathbf{S}_x \rangle$ using the standard estimator (averaged over the lattice volume) and the improved estimator. The measurements in the single-cluster method were done after each n_s updates. Table 1 shows the results for $C(x)$ at $x = 1, 20, 60, 120$, and for the susceptibility. To compare the performance the squared errors are shown, normalized to 20000 sweeps for each case.

Table 1. Spin-spin correlation function and susceptibility with different variants of the cluster algorithm (single-cluster, multi-cluster and 'hybrid'), measured with standard and improved estimators. The first line shows the data, the following lines the error squared, normalized to 20 ksweeps.

method	$C(1)$	$C(20)$	$C(60)$	$C(120)$	χ
H, imp.	0.68796(3)	0.1973(2)	0.0731(3)	0.0354(4)	3560(20)
S, st.	$2 \cdot 10^{-9}$	$5 \cdot 10^{-8}$	$1 \cdot 10^{-7}$	$2 \cdot 10^{-7}$	$6 \cdot 10^2$
M, st.	$1.6 \cdot 10^{-9}$	$3 \cdot 10^{-7}$	$9 \cdot 10^{-7}$	$1 \cdot 10^{-6}$	$2 \cdot 10^3$
H, st.	$1.3 \cdot 10^{-9}$	$9 \cdot 10^{-8}$	$3 \cdot 10^{-7}$	$4 \cdot 10^{-7}$	$7 \cdot 10^2$
S, imp.	$2 \cdot 10^{-5}$	$5 \cdot 10^{-6}$	$2 \cdot 10^{-6}$	$7 \cdot 10^{-7}$	$3 \cdot 10^3$
M, imp.	$1 \cdot 10^{-9}$	$1 \cdot 10^{-7}$	$3 \cdot 10^{-7}$	$4 \cdot 10^{-7}$	$1 \cdot 10^3$
H, imp.	$1 \cdot 10^{-9}$	$5 \cdot 10^{-8}$	$1 \cdot 10^{-7}$	$1.4 \cdot 10^{-7}$	$4 \cdot 10^2$

Comparing the standard estimators one sees that the single-cluster method updates all scales, except for the shortest distances better. The error squared for the hybrid method is somewhat larger because there we made ~ 1 single-cluster sweep (i.e. $n_s = 25$ single-cluster updates) followed by 3 multi-cluster updates with measurements. A striking observation is that for the single-cluster method the standard estimator produces a smaller error than the improved estimator – this is naturally connected with the noise discussed above. For the multi-cluster or hybrid method the corresponding improved estimator (which cancels this noise by taking a random orthogonal basis) obviously has smaller errors than the standard estimator, but the gain is about a factor of $2-3$, much smaller then predicted by the naive argument using the relations in (24). At $x = 60 \approx \xi$ the hybrid version gives an improvement in computer time of a factor 3 compared to the multi-cluster version and a factor of 20 compared to the single-cluster version with (noisy) improved estimator. It seems the best strategy is to use the single-cluster method with the *standard estimator* or the hybrid version with the improved estimator. Of course, at this correlation length all these versions of the cluster method are incomparably better than the local Metropolis update.

The cluster method still has many interesting applications for the $O(N)$ vector model [17–22]. As an example, let me mention here the determination of the interface tension in the Ising model by Hasenbusch [21]. He considered the Ising model on a 2d strip with different (periodic and anti-periodic) boundary conditions in the time direction. The boundary condition has been considered as a dynamical variable[3], $\epsilon = 1$ for periodic and $\epsilon = -1$ for anti-periodic b.c. The idea in [21] was that when a line of deleted bonds[4] cuts through the strip in the spatial direction one can flip (with probability 1) the region between this line and the boundary (at $t = 0$) flipping the sign of ϵ at the same time. The relative number of cases with $\epsilon = 1$ and -1 is related to the free energy of a kink, i.e. to the interface tension.

Unfortunately, the cluster algorithm does not work well for other models [24], at least in the most convenient version with independent clusters. The reason is always the same: the largest cluster tends to occupy the whole volume. To illustrate this, consider the Ising spin-glass,

$$E(S) = - \sum_{\langle i,j \rangle} J_{ij} S_i S_j \ , \tag{34}$$

where the random couplings can be positive or negative. In this case (when there are frustrated links) the clusters can grow through the boundaries separating the regions with $S = 1$ and $S = -1$, and for the couplings of interest one big cluster is formed. One can still give up the requirement of independent clusters and go back to (15). By decreasing $Q_l(S)$ one can make the clusters smaller – the price paid for this is that they start to interact and the acceptance probability may be extremely small for clusters of reasonable size. This possibility, however, has not been properly investigated yet.

3 Quantum Spin Systems

We shall consider the example of the 2d anti-ferromagnetic quantum Heisenberg model given by the Hamiltonian

$$H = J \sum_{x,\mu} \mathbf{S}_x \mathbf{S}_{x+\hat{\mu}} \ , \tag{35}$$

where $J > 0$ and $\mathbf{S}_x = \frac{1}{2}\boldsymbol{\sigma}_x$ is a spin operator ($\boldsymbol{\sigma}_x$ are Pauli matrices) at site x. This old model seems to describe the dynamics of the electron spins within the copper-oxygen planes of the La_2CuO_4 material. (Note that the first high-T_c super-conductor discovered was $La_{2-x}Ba_xCuO_4$ with the doping $x \approx 0.15$ [25].)

---

[3] A similar trick has been used in [23] for SU(3) gauge theory to determine the string tension.

[4] Sometimes it is more convenient to say that one originally puts bonds everywhere and then deletes them with probability $\bar{p}_l = 1 - p_l$.

To simulate a quantum spin system is much more difficult than a classical one because – as we shall see – the MC updates have to satisfy some constraints. Based on the work of Evertz, Lana and Marcu [26] who developed a loop cluster algorithm for vertex models, Wiese and Ying [27] worked out an analogous procedure for the Hamiltonian (35). I shall follow their derivation, with a few insignificant simplifications. In particular, I will consider the 1d case, and the multi-cluster version instead of the single-cluster one.

First we decompose the action into non-interacting sub-lattices, $H = H_1 + H_2$, where
$$H_1 = J \sum_n \mathbf{S}_{2n} \mathbf{S}_{2n+1} \; , \quad H_2 = J \sum_n \mathbf{S}_{2n-1} \mathbf{S}_{2n} \; , \tag{36}$$
and use the Suzuki–Trotter formula for the partition function:
$$Z = \mathrm{Tr} e^{-\beta H} = \lim_{N \to \infty} \mathrm{Tr} \left[e^{-\epsilon \beta H_1} e^{-\epsilon \beta H_2} \right]^N = \lim_{N \to \infty} \mathrm{Tr} \left(e^{-\epsilon \beta H_1} e^{-\epsilon \beta H_2} \cdots \right) \; , \tag{37}$$
where $\epsilon = 1/N$ determines the lattice spacing in the Euclidean time direction. By inserting a complete set of eigenstates $|+\rangle$ and $|-\rangle$ of σ_x^3 between the factors $\exp(-\epsilon \beta H_i)$ we obtain a classical system with Ising like variables $S(x,t) = \pm 1$ at each site of the 2d lattice. To see the structure of the corresponding contribution it is sufficient to consider a single interaction term in (36), $h = J \left(\mathbf{S}_a \mathbf{S}_b - \frac{1}{4} \right)$ where the term 1/4 is subtracted for convenience. Since $\mathbf{S}_a \mathbf{S}_b - \frac{1}{4} = \frac{1}{2}(\mathbf{S}_a + \mathbf{S}_b)^2 - 1$, the eigenvalues of h are $-J$ for the singlet state and 0 for the triplet state of the total angular momentum $\mathbf{S}_a + \mathbf{S}_b$. This gives the following transition amplitudes[5] for $A = \exp(-\epsilon \beta h)$:
$$\begin{aligned} \langle + + |A| + + \rangle &= \langle - - |A| - - \rangle \equiv w_1 = 1 \; , \\ \langle + - |A| + - \rangle &= \langle - + |A| - + \rangle \equiv w_2 = \tfrac{1}{2} \left(e^{\epsilon \beta J} + 1 \right) \; , \\ \langle + - |A| - + \rangle &= \langle - + |A| + - \rangle \equiv w_3 = \tfrac{1}{2} \left(e^{\epsilon \beta J} - 1 \right) \; . \end{aligned} \tag{38}$$
All other matrix elements are zero:
$$\langle + + |A| + - \rangle \equiv w_4 = 0 \; , \ldots \tag{39}$$
The separation $H = H_1 + H_2$ leads to a checker-board type interaction of the corresponding classical spin system – only the four spin on the corners of 'black' squares interact, producing the corresponding factor w_i in the Boltzmann factor. The partition function is given by
$$Z = \sum_{\{S\}} w_1^{n_1(S)} w_2^{n_2(S)} w_3^{n_3(S)} w_4^{n_4(S)} \cdots \tag{40}$$
where $n_i(S)$ is the number of 'black' plaquettes of type i. Clearly, only those configurations contribute for which $n_4(S) = n_5(S) = \ldots = 0$, i.e. the configurations should satisfy the corresponding constraints. This requirement causes a

---

[5] A direct calculation leads to a negative number, $-w_3$ for the third amplitude. By redefining the sign of the eigenvector $|-\rangle$ at every odd site one can make this amplitude positive without affecting the others.

serious problem for a local updating procedure since most steps are forbidden. In the loop cluster approach this problem is avoided – one builds closed loops of bonds on the configuration, and flips all spins along a loop. The new configuration automatically satisfies the constraints, and the bond probabilities are chosen to satisfy detailed balance.

The first step is to connect the sites of a 'black' square by bonds with some probabilities depending on the type of the configuration. The prescription is the following (taking the time axis to be the vertical one):

Table 2. Bond probabilities for different spin configurations on a plaquette.

type	bonds	probability	type	bonds	probability
1		1	2		$p = 1/w_2$
3		1	2		$p' = 1 - p = w_3/w_2$

It is easy to see that on a finite periodic lattice these bonds form a set of closed loops. Flipping all spins along a closed loop leads to a new admissible configuration (i.e. no forbidden 'black' plaquettes of type $i = 4, 5, \ldots$ appear). By flipping the spins on one of the bonds of a plaquette, the transition probabilities are $p(1 \to 2) = 1$, $p(2 \to 1) = 1/w_2 = w_1/w_2$, $p(2 \to 3) = w_3/w_2$, $p(3 \to 2) = 1$. These probabilities satisfy the relations

$$w_1 p(1 \to 2) = w_2 p(2 \to 1) \ , \quad w_2 p(2 \to 3) = w_3 p(3 \to 2) \ , \tag{41}$$

and consequently the condition of detailed balance for the equilibrium distribution corresponding to (40). This technique made it possible to determine the low energy effective parameters (as the spin wave velocity and spin stiffness) of the anti-ferromagnetic quantum Heisenberg model to a high precision [27].

In (37) one has to take the $\epsilon \to 0$ ($N \to \infty$) continuum limit in the time direction. Notice that the only non-deterministic choice is for plaquette of type 2 where the possibility of putting the link horizontally (along the spatial direction) is

$$p' = \frac{w_3}{w_2} = \frac{e^{\epsilon \beta J} - 1}{e^{\epsilon \beta J} + 1} = \frac{1}{2}\beta J \epsilon + O(\epsilon^2) \ . \tag{42}$$

One can now take a continuum time and reformulate the prescription into a continuum language: turning a path from the vertical into the horizontal direction in time interval dt equals $\frac{1}{2}\beta J dt$. Modifying the loop cluster algorithm this way Beard and Wiese [28] have shown that this observation allows the design of a loop cluster algorithm free from discretization error.

There are still many promising applications of the cluster algorithm to quantum systems but it is beyond our scope to discuss them here. I would like to mention only a recent suggestion by Galli [29] concerning the 'sign problem'.

4 Conclusion

Cluster algorithms work excellently for classical and quantum spin systems. The method of choosing the collective modes to be updated depends on the action and the actual configuration, and adjusts itself optimally. The Boltzmann weight (in the optimal case) is completely absorbed in the bond probabilities, hence the clusters can be updated independently. This leads to an effective update of large scale collective modes, to a strong reduction or even elimination of the critical slowing down, and to the possibility of introducing improved estimators with reduced variance.

References

[1] Madras, N., Sokal, A. D., *J. Stat. Phys.* **50** 109 (1988) .
[2] Adler, S. L., *Phys. Rev.* **D 23** 2901 (1981); *Phys. Rev. Lett.* **66** 1807 (1991).
[3] Edwards, R. G., Ferreira, S. J., Goodman J., Sokal, A. D., *Nucl. Phys.* **B 380** 621 (1992).
[4] Swendsen, R. H., Wang, J. S., *Phys. Rev. Lett.* **58** 86 (1987).
[5] Kasteleyn, P. W., Fortuin, C. M., *J. Phys. Soc. Jpn. Suppl.* **26s** 11 (1969); Fortuin, C. M., Kasteleyn, P. W., *Physica* (Utrecht) **57** 536 (1972).
[6] Niedermayer, F., *Phys. Rev. Lett.* **61** 2026 (1988).
[7] Krauth, W., this proceedings.
[8] Wolff, U., *Phys. Rev. Lett.* **62** 361 (1989).
[9] Wolff, U., *Nucl. Phys.* **B 322** 579 (1989).
[10] Wolff, U., *Phys. Lett.* **B 222** 473 (1989).
[11] Edwards, R. G., Sokal, A. D., *Phys. Rev.* **D 40** 1374 (1989).
[12] Symanzik, K., in *Recent developments in gauge theories*, Cargese, ed. G. 't Hooft et al. (Plenum, New York, 1980).
[13] Hasenfratz, P., Niedermayer, F., *Nucl. Phys.* **B 337** 233 (1990).
[14] Dimitrovic, I., Hasenfratz, P., Nager J., Niedermayer, F., *Nucl. Phys.* **B 350** 893 (1991).
[15] Hasenfratz, P., Niedermayer, F., *Nucl. Phys.* **B 414** 785 (1994).
[16] Niedermayer, F., *Physics Letters* **B 237** 473 (1990).
[17] Janke, W., *Phys. Lett.* **A 148** 306 (1990).
[18] Holm C., Janke, W., *Nucl.Phys.* **B** (Proc.Suppl.) **30** 846 (1993).
[19] Creutz, M., *Phys. Rev. Lett.* **69** 1002 (1992).
[20] Janke W., Kappler, S., *Phys. Rev. Lett.* **74** 212 (1995).

[21] Hasenbusch, M., *J. Phys.* **I 3** 753 (1993).
[22] Kerler, W., *Phys. Rev.* **D 48** 902 (1993).
[23] Hasenfratz, A., Hasenfratz P., Niedermayer, F., *Nucl. Phys.* **B 329** 739 (1990).
[24] Sokal, A. D., *Nucl. Phys.* **B** (Proc. Suppl.) **20** 55 (1991).
[25] Bednorz, J. G, Müller, K. A., *Z. Phys.* **B 64** 189 (1986).
[26] Evertz, H. G., Lana, G., Marcu, M., *Phys. Rev. Lett.* **70** 875 (1993).
[27] Wiese, U.-J., Ying, H.-Y., *Z. Phys.* **B 93** 147 (1994).
[28] Beard, B. B., Wiese, U.-J., cond-mat/9602164.
[29] Galli, A., hep-lat/9605007, 9605026.

Optimized Monte Carlo Methods

Enzo Marinari

Dipartimento di Fisica, Università di Cagliari, via Ospedale 72, 09100 Cagliari (Italy)

Abstract. I discuss optimized data analysis and Monte Carlo methods. Reweighting methods are discussed through examples, such as Lee-Yang zeroes in the Ising model and the absence of deconfinement in QCD. Reweighted data analysis and multi-histogramming are also discussed. I introduce simulated tempering, and, as an example, its application to the random field Ising model. I illustrate parallel tempering, and discuss some crucial technical details such as thermalization and volume scaling. I give a general perspective by discussing umbrella methods and the multicanonical approach.

1 Introduction

In the following I will give an introduction to optimized Monte Carlo methods and data analysis approaches. We shall see that the two issues are closely interconnected, and we will try to understand why. I will try to keep the same style one has while lecturing, trying to explain all important points in some detail. Even the figures will be mostly copied from my transparencies: I hope that this will, at least partially, fulfill the goal I have in mind.

2 Reweighting Methods

Reweighting methods are based on a simple, basic idea: when you run a numerical simulation at a given value of the inverse temperature β and you measure some set of observable quantities $O^{(\alpha)}$ (including the internal energy of the system) you learn far more than simply the value of

$$\langle O^{(\alpha)} \rangle , \qquad (1)$$

and the mesure of your ignorance about it (the statistical error). Expectation values of the observable quantities at β turn out to be only a small part of the information you are gathering (if you store the right numbers!).

The partition function of our statistical system at β can be expressed as the integral over the energy levels

$$Z_\beta = \int dE \ N(E) \ e^{-\beta E} , \qquad (2)$$

where $N(E)$ is the energy density. *An accurate numerical simulation at β gives you information about $N(E)$*, and this information can be used in many ways, as we will discuss in the following. By an accurate numerical simulation we mean here that in order to make the information about $N(E)$ meaningful we need a large sample, and that the problem of controlling the statistical and systematic errors is non-trivial. This is also the path to the definition of improved Monte Carlo methods like *tempering*, that we will discuss in the following.

In the following we will discuss reweighting techniques (also as an introduction, as we said, to the improved Monte Carlo method, to make the logical path that leads us there clear). Following the discussion of Falconi et al. [1], Marinari [2], we will first discuss the simple Ising model and the Lee-Yang theorem in (2.1) and then the existence of a phase transition in a 4 dimensional $SU(2)$ lattice gauge theory in (2.2). By doing that we will give a very schematic definition of a lattice gauge theory. We will next discuss the use of this approach for improving the quality of the analysis of numerical data. We will introduce histogramming in (2.3) (this is a classical development, based on classical work in molecular dynamical simulations (by Salsburg et al. [3], Chestnut and Salsburg [4], Mc Donald and Singer [5],) and on the more recent work contained in [1], [2] and [6]), and the work on multi-histogramming of Ferrenberg and Swendsen [7] in (2.4).

So, in order to summarize again the physics side of the next section, we will start by discussing Lee-Yang zeroes in the $3d$ Ising model, and clarifying a few crucial issues about phase transitions. We will discuss how to compute critical exponents from there. Then we will discuss the analysis of zeroes in $SU(2)$ lattice gauge theory and we will see how that helps in establishing that the *confinement of quarks in colorless particles survives in the continuum limit*. At last we will give details about data analysis. We will also try to clarify the path that will eventually lead this technique to be promoted, from a data analysis tool, to a (sometimes very effective) simulation method.

2.1 Lee-Yang zeroes

Someone interested in the numerical study of critical phenomena should always consider as fundamental the fact that phase transitions only exist in the infinite volume limit. In a finite volume (i.e. inside our computer) there are no phase transitions. Let us start by clarifying this point a bit.

We are working in the complex β (or T) plane. We consider a compact configuration space (spin variables cannot diverge: the Ising case with ± 1 values is a very good example, as is also a compact $SU(N)$ gauge theory with $SU(N)$ matrices) but what we will discuss can also be proved under far more general assumptions. The absolute value of the partition function Z_β is limited from above by

$$|Z_\beta| \equiv \left| \int \{dC\} e^{-\beta H(\{C\})} \right| \leq V_C e^{|\beta||H|} , \qquad (3)$$

where the integral is over the configuration space dC, $V_c \equiv \int \{dC\}$ is the volume of the configuration space and by $|H|$ we denote the maximum value $H(\{C\})$

can assume when considering all possible configurations:

$$|H| \equiv \max_{\{C\}}\Big(H(\{C\})\Big) \ . \quad (4)$$

This is also true for complex β values (when the spin variables can only take discrete values Z is a linear combination of exponentials).

What are the properties of the partition function Z_β? A reasonable Z_β, which is supposed to describe a physical system, is an analytic function in the $\text{Re}(\beta) > 0$ half of the complex plane. And what happens to the free energy, $-\frac{1}{\beta V}\log Z_\beta$? If Z is an analytic function $\log Z_\beta$ can be singular only where $Z_\beta = 0$. That makes the role of the *zeroes of the partition function* clear. For T (and the field h) belonging to R^+ Z_β is a sum of positive contributions, and cannot have zeroes in the finite volume V. But, as Fig. 1 illustrates, in the $V \to \infty$ limit zeroes that are located for finite V at complex values of $T_0 \in C$ can approach the real T axis, generating a non-analyticity of the free energy.

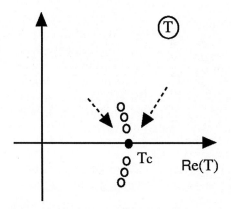

Fig. 1. The complex T plane. Zeroes at finite volume (empty dots) pinch the real axis at T_c (filled dot) in the infinite volume limit.

The same kind of reasoning can be applied to the behavior in a magnetic field h: here for the Ising model the *Lee-Yang* theorem holds: *The zeroes of the partition function are located on the imaginary h-axis, or on the unit circle of the complex activity plane*. In a finite volume there are a finite number of zeroes, and in the infinite volume limit the zeroes condense on part of the unit circle.

There are no theorems about constraining the zeroes in the complex β-plane. The main theoretical results on this issue had been obtained in [8]:

- The complex zeroes close to the (to be) critical point accumulate on curves;
- In $2d$ they cross the real axis at T_c at a right angle.

We will assume that the same situation holds (with a generic crossing angle $\pi - \phi$) in $d > 2$. We expect two lines of zeroes in the upper and lower positive T complex plane (symmetric because of the reality properties of the partition function Z), that pinch the real T axis at T_c. The singular part of the free energy above and below the transition has the form

$$f^{(\pm)}_{\text{singular}} \simeq A^{(\pm)} \left(\frac{T - T_c}{T_c} \right)^{2-\alpha} , \tag{5}$$

where α is the usual critical exponent and $A^{(+)}$ and $A^{(-)}$ are the specific heat critical amplitudes. Matching the lines of zeroes gives the condition

$$\tan\left((2-\alpha)\phi\right) = \frac{\cos(\pi\alpha) - \frac{A^{(-)}}{A^{(+)}}}{\sin(\pi\alpha)} . \tag{6}$$

In $2d$ $\phi = \pi/2$ gives that $A^{(-)} = A^{(+)}$, as it should.

So, we have scaling laws for the position of complex zeroes of the partition function. Finite size dependence can be derived in the usual way, and we will be able to try a numerical experiment to determine the critical behavior.

The numerical simulation will be based, as we said, on the fact that from a Monte Carlo simulation at a fixed β value we can gather information about other values of β (even complex values). Running simulations directly at complex β values is far from straightforward, and we will find this approach quite useful: the same approach will lead us to introduce very powerful Monte Carlo methods. Here we will use the method to compute zeroes of $Z_\beta^{(V)}$ at $\beta_0 = (\text{Re}(\beta_0), \text{Im}(\beta_0))$, with small $\text{Im}(\beta_0)$ (this is the region that is the most interesting from the scaling point of view and, luckily enough, is also the one that we can access by numerical methods). When we apply the method to real, small β increments we will get a useful tool for data analysis.

We can start being specific (following [2]), and consider the $3d$ Ising model, with spin variables $\sigma_i = \pm 1$, i a $3d$ label of a lattice site, and an action S

$$S = -\sum{}'(\sigma_i \sigma_j - 1) , \tag{7}$$

where the primed sum runs on first neighbor spin couples on a simple cubic lattice. The partition function Z_β is written as

$$Z_\beta = \sum_{\{C\}} e^{-\beta S(\{C\})} . \tag{8}$$

At the time of this work an exact solution by enumeration had been obtained for lattices of size up to 4^3 by Pearson [9], one must count $O(10^{19})$ configurations in this case! Already a lattice of 5^3 sites cannot be exactly enumerated with today's computers. As we said our statistical technique will be based on the fact that we can express the partition function as a sum over the energy levels of the system:

$$Z(\beta) = \sum_E N_E e^{-\beta E} , \tag{9}$$

where in our case $E = 0, \ldots, 3L^3$. The instructions are: run your Monte Carlo simulation at $\tilde{\beta}$, and record the normalized energy distribution function $F_E(\tilde{\beta})$ (this is the number of configurations you find at each energy value, normalized to one). One has

$$F_E(\tilde{\beta}) = \frac{f_E(\tilde{\beta})}{Z(\tilde{\beta})} = \frac{N_E \, e^{-\tilde{\beta}E}}{Z(\tilde{\beta})}$$

$$\sum_E F_E(\tilde{\beta}) = 1 \;. \tag{10}$$

Now if we compare two different β values (we have run the simulation at $\tilde{\beta}$ and we are interested in results at β) we see from (10) that

$$N_E = F_E(\beta) Z(\beta) e^{\beta E} = F_E(\tilde{\beta}) Z(\tilde{\beta}) e^{\tilde{\beta}E} \;. \tag{11}$$

We can already see that if we stop here, assume that we are dealing with two real β values, and we use our best numerical estimate for the partition function, we get the reweighting formula

$$F_E(\beta) = F_E(\tilde{\beta}) \frac{e^{(\tilde{\beta}-\beta)E}}{\sum_E F_E(\tilde{\beta}) e^{(\tilde{\beta}-\beta)E}} \;. \tag{12}$$

As we will see better in the following, and we are only anticipating here, from the simulation at $\tilde{\beta}$ we can get expectation values at β (if β is close enough to $\tilde{\beta}$, and the statistical accuracy is good enough not to make the exponential suppression kill the signal). But at the moment let us go back to the complex zeroes, and rewrite (11) as

$$F_E(\beta) = F_E(\tilde{\beta}) \, e^{(\tilde{\beta}-\beta)E} \frac{Z(\tilde{\beta})}{Z(\beta)} \;. \tag{13}$$

Summing over energies and using the normalization condition of (10) we find

$$\frac{Z(\beta)}{Z(\tilde{\beta})} = \sum_E F_E(\tilde{\beta}) \, e^{(\tilde{\beta}-\beta)E} \;. \tag{14}$$

So, we are running a Monte Carlo simulation at $\tilde{\beta}$, and we obtain a numerical estimate for $F_E(\tilde{\beta})$. We want to obtain information about the analytic structure of Z_β at $\beta \equiv \eta + i\xi$. The exponential factor in (14) will give, for complex β, two contributions: the first will be an oscillating factor, that is non-zero for $\text{Im}(\beta) \neq 0$,

$$\cos(E\xi) + i \sin(E\xi) \tag{15}$$

(we have to remember that E here is a number of the order of the volume, not of order one). The other contribution is the exponential damping

$$e^{-(\eta-\tilde{\beta})E} \;: \tag{16}$$

since $E = O(V)$ the damping is severe. Since we are looking for zeroes and we know we cannot get zeroes on the real axis, it is a good idea to compute

$$\frac{Z_\beta}{Z_{\text{Re}\beta}} = \frac{Z_\beta}{Z_{\tilde\beta}} \left\{ \frac{Z_{\text{Re}\beta}}{Z_{\tilde\beta}} \right\}^{-1} \qquad (17)$$

$$= \frac{\sum_E F_E(\tilde\beta) e^{-(\eta-\tilde\beta)E} \left(\cos(E\xi) + i\sin(E\xi)\right)}{\sum_E F_E(\tilde\beta) e^{-(\eta-\tilde\beta)E}}.$$

An easy way for a numerical search of zeroes of (17) is to look for minima of

$$\left|\frac{Z_\beta}{Z_{\text{Re}\beta}}\right|^2 = \qquad (18)$$

$$\frac{\left(\sum_E F_E(\tilde\beta) e^{-(\eta-\tilde\beta)E} \cos(E\xi)\right)^2 + \left(\sum_E F_E(\tilde\beta) e^{-(\eta-\tilde\beta)E} \sin(E\xi)\right)^2}{\left(\sum_E F_E(\tilde\beta) e^{-(\eta-\tilde\beta)E}\right)^2}.$$

The numerical simulations in [2] were using volumes from $N^3 = 4^3$ (in order to check the exact solution) to 8^3. $\tilde\beta$ was chosen as close as possible to the actual location of the zero in order to minimize the exponential damping.

We have already said that $\cos(E\xi) = \cos(eV\xi)$ (where the energy density e of order 1) is rapidly oscillating, and makes it impossible to compute the location of zeroes with a large imaginary part. One nice fact is that by finite size scaling we expect that the distance δ_N of the first zero on a lattice of linear size N scales as

$$\delta_N \simeq N^{-\frac{1}{\nu}}. \qquad (19)$$

Equation (19) can be used to estimate ν from the rate at which zeroes approach the real axis. It also tells us that the oscillations due to the cosine term do not increase like N_d, but rather like $L^{d-\frac{1}{\nu}}$: an exponent of 1.4 instead of 3 for the 3d Ising model. This helps.

In Fig. 2 we show the scaling of the distance of the first zero from the real axis (one can do the same for farther zeroes, but the precision is smaller). We use here the variable $u \equiv e^{-4\beta}$. We denote by u_N^i the position of the i-th zero on a lattice of linear size N, and plot u_N^0 versus N in double log scale (the figure here is not precise, and it is only meant as a graphical reconstruction of the data: the reader interested in raw numbers should consult directly [2]).

The remarkable linearity of the plot displays a good scaling behavior already for small lattices. From these data (numerical archeology today, but we are discussing the method, not the numbers!) one gets the reasonable estimate $\nu \simeq .620 \pm .010$ (the best estimate for ν at the time of this simulation was the far superior $\nu \simeq .631 \pm .001$ by Le Guillou and Zinn-Justin [10]). But with this method, for example, we can get a very straightforward measurement of the ratio of the critical amplitudes (a quantity that is not easy to obtain in the series expansion approach). We measure, as we have discussed before, the angle

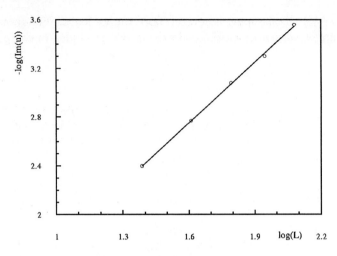

Fig. 2. $-\log \operatorname{Im} u_N^0$ versus $\log L$.

ϕ at which the complex zeroes depart from the real u axis in the infinite volume limit. One has

$$\tan((2-\alpha)\phi) = \frac{\cos(\pi\alpha) - \frac{A^{(+)}}{A^{(-)}}}{\sin(\pi\alpha)}. \tag{20}$$

So, for example, in [2] we got an angle of 55.3 ± 1.5 degrees, and an amplitude ratio of 0.45 ± 0.07. This is an accurate and reasonable result. The method works[1]!

2.2 Complex zeroes in a non-Abelian four dimensional lattice gauge theory

In the previous section we have described the method one can use to locate complex zeroes of the partition function. We will now discuss the physical problem for which this technique was first introduced in [1]. I am basically taking this chance to give a ten lines crash course in lattice gauge theories, LGT (that for our purposes will mainly be a sort of fancy statistical mechanics, constructed by exploiting a powerful symmetry). We will discuss locating complex zeroes in a non-abelian $4d$ LGT, i.e. one of the ways of getting numerical evidence to show that there is no deconfining phase transition in the infinite volume limit. The interested reader can find relevant background material in Rothe's [12] or Montvay and Münster's [13] book, in the classical lecture notes by Kogut [14], or in the Les Houches notes by Parisi [15].

In a lattice gauge theory variables live on *links* (as opposite to sites for a normal statistical mechanics) of a d-dimensional lattice (simple cubic, for simplicity,

---

[1] Obviously there are more recent simulations that follow these lines and they are more precise: see for example [11], and references therein.

in the following), and we will label them by $U_\mu(n)$, where n is a d-dimensional site label, and μ denotes one of the lattice directions (we show the link variable in Fig. 3a). Different gauge theories are characterized by different kind of variables. In the case of quantum chromodynamics, the theory of strong interaction of elementary particles, they are $SU(3)$ matrices (here we will consider a simpler theory with many similar features, the one of $SU(2)$ 2×2 matrices). In the case of an $SU(M)$ gauge group U is an $M \times M$ matrix with $U \cdot U^\dagger = 1$, and $\det(U) = 1$.

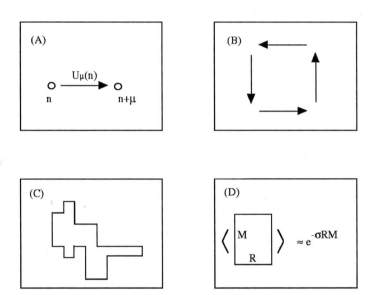

Fig. 3. From gauge variables to Wilson loops. See text.

The Boltzmann equilibrium probability distribution can be written as

$$P_B \simeq e^{\beta \sum_P U_P} , \qquad (21)$$

where the sum runs over all *plaquettes* (the smallest closed circuits, see Fig. 3b) of the lattice, and

$$U_P \equiv U_\mu(n) U_\nu(n + \hat{\mu}) U_\mu^\dagger(n + \hat{\nu}) U_\nu^\dagger(n) , \qquad (22)$$

i.e. one takes on each elementary plaquette the product of ordered link matrices (by defining $U_\mu(n) = U_{-\mu}^\dagger(n + \hat{\mu})$).

This theory has a dramatically large invariance, known as *gauge invariance*: if we arbitrarily pick a group element $g(n)$, that we can choose independently on each site, and we transform all the link variables under

$$U_\mu(n) \to g(n) U_\mu(n) g^\dagger(n + \hat{\mu}) , \qquad (23)$$

the action of the theory does not change (and neither do any observable quantities, i.e. products of links on closed loops, see Fig. 3c). In the Ising model the crucial Z_2 symmetry is only a *global* symmetry: the theory is symmetric under inversion of all $\sigma(n)$ variables. Here, on the other hand, we have the right to select an independent frame rotation at each lattice point, and the theory does not change. Such a gauge symmetry is exact in the lattice theory. Preservation of gauge symmetry is the crucial step of the Wilson approach to LGT: in the same way in which the 2d Ising model Onsager solution is described by Majorana fermions at the critical point, where the correlation length becomes infinite and details of the lattice structure are forgotten, the critical limit of lattice QCD is the usual continuum QCD, the theory of interacting quarks and gluons. The fact that continuum gauge invariance is exactly preserved on the lattice (as opposed to Lorentz invariance, for example, which is obviously broken by lattice discretization) is a crucial point of the approach.

We have said, and we will not go into details, that products of link variables on closed loops are the observable quantities.

We also remark that the inverse temperature β that appears in the Boltzmann distribution (21) is connected to the coupling constant of the continuum gauge theory one recovers in the limit of small lattice spacing:

$$\beta_{StM} \simeq g_{GT}^{-2}, \quad T_{StM} \simeq g_{GT}^{2}. \tag{24}$$

The theory has a phase transition at $T = 0$ (here $g \to 0$), where the correlation length diverges (exponentially in $\frac{1}{T^2}$). As usual in this continuum limit the lattice structure becomes irrelevant.

For high values of T it is easy to prove the relation we have depicted in Fig. 3d: a Wilson loop (the product of oriented link variables over a closed loop) of large size $R \cdot M$ decays as the exponential of minus the *string tension* σ times the area of the loop in the confined region. If quarks are confined and cannot be separated out from color singlet states (the physical particles, mesons and baryons) we have an area decay of large Wilson loops. So, the nice part is that, as we have said, it is easy to prove that lattice QCD is confined in the high T region (where the theory is very different from the continuum theory). The bad part is that one can prove the above statement for all LGT, including the lattice QED. Since electrons are known to be free in the continuum theory, in this case something will have to happen. What happens for example in the case of QED is that a finite T phase transition separates two regions, the confined, unphysical one, and the deconfined physical theory. One has to show that the same does not happen in Lattice QCD, and that the theory does not undergo a phase transition that would destroy confinement.

The purpose of our analysis of lattice zeroes was to study this problem. The technique is exactly the same as what we have described in the former section. In Fig. 4 we draw curves with

$$\text{Re}\left(\frac{Z(\eta + i\xi)}{Z(\eta)}\right) = 0, \tag{25}$$

and curves with
$$\text{Im}\left(\frac{Z(\eta+i\xi)}{Z(\eta)}\right) = 0 \qquad (26)$$

(for the exact curves the reader should consult the original reference, [1]). It is clear from the figure that one can determine the zero with good precision. Since one finds a zero quite far from the real β axis, and it does not approach the real axis for larger lattice sizes, one does not expect a real phase transition, but is measuring a transient phenomenon that is irrelevant as far as the real continuum limit is concerned. It is clear that the evidence we have discussed here is the same as what one exploits when using finite size scaling techniques, looking for example at the behavior of a peak of the specific heat. It is interesting that one can directly study the position of complex zeroes of the partition function.

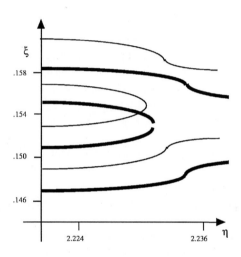

Fig. 4. Lines of zeroes of the imaginary part (thin curves) and of the real part (thick curves) of the ratio of partition functions.

2.3 Data analysis

In equation (12) we have already seen that our approach can be used to deduce information at β after running a simulation at $\tilde{\beta}$. We can generalize (12) by noticing that we can also sample magnetizations m (so we can reconstruct all moments of m). After each measurement we write down the energy and the magnetization of the configuration (that we assume to be at thermal equilibrium). We have that

$$F_{E,m}(\beta,h) = F_{E,m}(\tilde{\beta},\tilde{h}) \frac{e^{(\tilde{\beta}-\beta)E+(\tilde{h}-h)m}}{\sum_{\tilde{E},\tilde{m}} F_{\tilde{E},\tilde{m}}(\tilde{\beta},\tilde{h}) e^{(\tilde{\beta}-\beta)E+(\tilde{h}-h)m}} \ . \qquad (27)$$

In [6] one finds a very nice picture showing how well the method can work for example for the case of the $2d$ Ising model.

The method we have described here is very useful when one wants to measure the finite size scaling behavior of physical observable quantities. Let us consider for example the specific heat C_V. The maximum value C_V takes on a finite lattice of linear size L, $C_V^{\max}(L)$ scales at a critical point as

$$C_V^{\max}(L) \simeq L^{\frac{\alpha}{\nu}} . \tag{28}$$

The main problem in MC simulation is that we only measure for a discrete set of values of the temperature T. We do not know a priori at which value of T on a given lattice the specific heat takes its maximum value, and such a T value depends on L:

$$T^{\max}(L) \mid C_V(T^{\max}(L), L) \equiv C_V^{\max}(L) \text{ depends on } L . \tag{29}$$

It can be difficult to find the correct value of $T^{\max}(L)$: it is a delicate fine tuning process that has to be repeated for each different L value. An incorrect determination of $T^{\max}(L)$ can generate a very misleading effect. Let us look at Fig. 5. The empty dots represent measurements of the specific heat taken with a linear size L, at the set of T values which appear in the figure. The filled dots are measurements on a lattice of larger linear size L', taken at the same T values. One would assume that the finite size scaling behavior is given by the scaling of the two points with the highest value of C_V. But maybe the real maximum, on the larger lattice, is where the triangle is: there the scaling could be very different from the one we found on the points we measured, but sadly we did not measure on this point. We want to stress that this effect creates real practical problems in numerical simulations.

The pattern of data analysis we have discussed here solves this important problem. Statistical error can be kept under control (for example by using jackknife and binning techniques), and numerical studies of scaling become (without further computational expenses, since we already had the information) a real quantitative tool.

2.4 Multi-histogramming

The idea of reweighting can be pursued further. Ferrenberg and Swendsen [7] introduced *multi-histogramming*. The crucial step is to realize that you can patch data from different simulations at different β_k values to reconstruct all of the relevant β region.

So, we have to sum up histograms. The delicate point is how to sum them up, i.e. how to determine the weights g_n to use in constructing linear combinations of the different entries. The method discussed in [7] is to determine the weights by minimizing the statistical uncertainty over the final estimate for $P_\beta(E)$. Let us call $N_k(E)$ the data histogram for the k-th Monte Carlo run, at β_k. Let us define $\theta_k \equiv 1 + 2\tau_k$, where τ_k is the estimated correlation time at β_k, and n_k

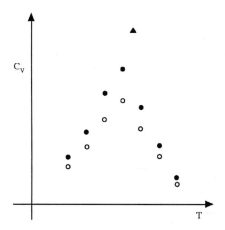

Fig. 5. The specific heat versus T for a typical finite size scaling study suffering from troubles that can be cured by reweighting. Empty dots are for the smaller lattice, filled dots for the larger lattice, and the triangle for the point we did not measure and we should have measured.

the number of sweeps used in the k-th Monte Carlo run. One finds that the parameters $\{g_n\}$ can be determined self-consistently by iteration from

$$P_\beta(E, \beta_m) = \frac{\sum_{k=1}^{R} N_k(E) \, e^{\beta_m E} \, \theta_k^{-1}}{\sum_{j=1}^{R} n_j \, e^{\beta_j E - g_j} \, \theta_j^{-1}} \; ;$$

$$e^{g_m} = \sum_E P_\beta(E, \beta_m) \; . \tag{30}$$

We have denoted by R the number of Monte Carlo runs. The method works well, and today it can be considered as a standard tool of analysis.

3 Improved Monte Carlo Methods

Here we will discuss improved Monte Carlo methods. We will mainly be talking about *tempering*, introduced by Marinari and Parisi [16] (see section (3.1)), where you add a second Markov chain to the usual Metropolis chain: you let β vary, trying in this way to make it easier for the system to move across deep free energy valleys separated by high free energy barriers. We will also try to discuss general issues such as *thermalization*, which are of crucial importance already when discussing simple Monte Carlo methods, and which turn out to be even more delicate issues here.

Ultimately one is looking for a very effective, very simple simulation method. Somehow when you start the numerical study of a statistical system you work at two different levels. At first you run a quick and not very clean Monte Carlo, to gain a preliminary understanding[2]. Only afterwards do you set up complex simulation procedures, data analysis and error determination. It would be nice if the first phase we have just described could already be based on something more powerful that the usual Metropolis approach: I hope the reader will be convinced that maybe the *parallel tempering* approach (see section (3.3)) is the good candidate for this role. In parallel tempering there are no parameters to be tuned, no difficult choices to be made (except for the selection of a set of T values that can be done with a large amount of liberty): it looks like the right thing.

3.1 Simulated tempering

We will introduce here *simulated tempering*, [16], an improved Monte Carlo method that turns out for example to be very effective for simulating spin glasses (for further studies and applications of tempering see [17–21]). Later on we will discuss *parallel tempering*, [22–25], that turns out to be a better and simpler method (having both these advantages at the same time is quite a rare and appreciated feature). So we will now discuss some complex matter that we will eventually not need in the practical implementation of the method but that will help in understanding some of the physical mechanisms that govern the scheme. These mechanisms are also shared by the most promising parallel tempering schemes, where parameter tuning is not needed.

Simulated tempering is a global optimization method: it can be seen as an annealing with a built-in scheduling. This can be of great practical importance, since setting up the schedule is one of the most difficult and time consuming parts of an annealing simulation. Tempering is very similar to annealing, but after the initial thermalization period the field configuration is always at thermal equilibrium at one of the allowed β values. This phrase, which is a bit mysterious at this point, is important, and will be clarified in the following.

There are many related techniques, strictly connected methods and necessary introductory material. First of all one needs to know the basics of Monte Carlo methods (see for example the lectures in this book by Krauth [26]). The theory of multiple Markov chains is the mathematical basis one needs to clarify the theoretical aspects of the method [22–25].

Strictly related to tempering are the *scaling* approaches based on *umbrella sampling*, [27–30]: we have already discussed the issue when illustrating data analysis reconstruction. It is important to notice that many of the ideas we are now applying to numerical statistical mechanics and Euclidean statistical field theory had been elaborated many years ago in the context of the physics of liquids and of molecular dynamics.

The *multicanonical approaches* by [31–34] are also closely related to tempering, and we will discuss them in the following. Multicanonical methods are more

---

[2] The phrase typically used by G. Parisi to describe this approach is *Il buon giorno si vede dal mattino*, that I would translate in English as *Early in the morning you can already tell a good day from a bad day*.

ambitious and in many situations potentially more powerful than tempering (but they are a bit more complex): they can deal with first order phase transitions, which is not something you want to do with tempering (that works well for continuous phase transitions). We also note that different kinds of tempering-like approaches have been proposed, for example in [35].

As we have discussed from the start of these notes tempering is built on reconstruction methods, [1], [2], [6], [7]: if we can use data at $\tilde{\beta}$ for learning about expectation values at β maybe we can also use this information for speeding up the dynamics itself (again, see [26] for an introduction to Monte Carlo simulations and Markov chains). When you run a long simulation you could learn much more than you believe (or much less, but we will discuss that when talking about thermalization and correlation times).

Let us start with describing *simulated tempering*. We want to equilibrate our statistical system with respect to the Boltzmann distribution

$$P(\{\sigma\}) \simeq e^{-\beta S(\{\sigma\})} \, . \tag{31}$$

We choose a *new* probability distribution, with an enlarged number of variables:

$$\tilde{P}(\{\sigma\}, \{\Sigma\}) \, , \tag{32}$$

such that, for each given choice of the $\{\Sigma\}$, $\tilde{P}$ is a Boltzmann distribution with some given choice of β. We will allow β to become a *dynamical variable*:

$$\{\sigma\} \to (\{\sigma\}, \{\beta_\alpha\}) \, , \tag{33}$$

$\alpha = 1, \ldots, A$. We have allowed A fixed values for the β_α variables.

The mechanism we are introducing is very simple: the new equilibrium probability distribution is

$$P_{eq}(\{\sigma\}, \{\beta_\alpha\}) \simeq e^{-\beta_\alpha H(\{\sigma\}) + g_\alpha} \, , \tag{34}$$

where H is the original Hamiltonian of the problem. The g_α are constant quantities, whose meaning will be elucidated in the following. They have to be fine tuned before running the equilibrium sampling, and they only depend on the value of α (only one g_α is allowed for each β_α value).

The probability of finding a given value of α (i.e. the probability for a given β_α value to occur during the run) is:

$$P_\alpha = z_\alpha e^{g_\alpha} \equiv e^{g_\alpha - \beta_\alpha f_\alpha} \, , \tag{35}$$

where z_α is the partition function at fixed β_α,

$$\int \{d\sigma\} e^{-\beta_\alpha H(\{\sigma\})} \, , \tag{36}$$

and f_α is the free energy of the system with fixed β_α. (35) shows that the free parameters g_α are related to the free energy of the system. In order to make the probability of finding the system in each value of β the same (i.e. the system to visit all allowed β values with the same frequency) we need to set

$$g_\alpha = \beta_\alpha f_\alpha \ . \tag{37}$$

Since we do not know the f_α that will amount to use for the g's the best available estimate for f. We will discuss in the following how to produce a good guess for the g_α that can also be improved systematically. Since parallel tempering will eventually not need any kind of parameters to be fine tuned we will not go into details about this issue (explaining what the g's are in simple tempering is useful for reaching a better physical understanding of both tempering and parallel tempering).

If we are mainly interested in studying the system at $T = \tilde{T}$ we will allow for a set of T_α, for example at constant distance, $T_0 \equiv \tilde{T}$, $T_1 \equiv (\tilde{T} + \delta)$, $T_2 \equiv (\tilde{T} + 2\delta)$, ... (see Fig. 6). We will discuss how to select optimal values for δ (this has to be done also in parallel tempering, but one does not need fine tuning). After the runs if you want you can use reconstruction schemes to use all of the information contained in all samples, but in this kind of approach this is typically not necessary (since we are looking for thermalized configurations in a complex, low temperature phase, and the main information is typically contained in the T_0 data).

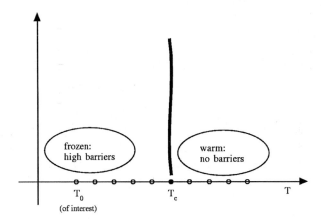

Fig. 6. The choice of the T values to be used in the tempered simulation of a complex system.

The main physical ideas are the following:

1. The system is frozen at low T. High energy barriers separate different free energy valleys.
2. When the system warms up during tempering the free energy barriers get smaller. They disappear when the system crosses T_c.

3. When it cools down again it will probably explore a different local minimum.

The method seems to work nicely for second order like phase transitions: for that to happen you need the broken state to be conformationally similar to the high T one.

We will now analyze in detail a full updating sweep, in order to make the procedure clear.

1) Sweep the full lattice, maybe s times ($s > 1$ could help) and run a completely normal Monte Carlo (Metropolis, or cluster, or over-relaxed or what you like best) update for the $\{\sigma\}$ variables at *fixed* β_α.

2) Propose the update

$$\beta_\alpha \to \beta_{\alpha'} , \tag{38}$$

where $\alpha' = \alpha \pm 1$ with probability $\frac{1}{2}$ (at $\alpha = 1$ and $\alpha = A$ we may or may not decide to move with probability $\frac{1}{2}$, since we are not interested in detailed balance in β space: see [26] about the importance of rejected moves).

We accept the update with normal Metropolis weight. Here the factor in the exponential only changes because of the change in β, since the configuration energy does not change:

$$\Delta S_{\text{tempering}} = (\beta_{\alpha'} - \beta_\alpha)H(\{\sigma\}) + (g_{\alpha'} - g_\alpha) , \tag{39}$$

where both the H and the $(g_{\alpha'} - g_\alpha)$ terms are constant. If $\Delta S \le 0$ we accept the β update, if $\Delta S > 0$ we accept it if a random number uniformly distributed in $(0,1)$ is smaller than $e^{-\Delta S}$.

A good first guess for the g_α can be deduced from (39) (then, as we said, the g's can be systematically improved). We can take the g such that in average ΔS is balanced:

$$(g_{\alpha'} - g_\alpha) = -(\beta_{\alpha'} - \beta_\alpha)\frac{\langle H \rangle_{\beta_\alpha} + \langle H \rangle_{\beta_{\alpha'}}}{2} . \tag{40}$$

This holds at first order in $\delta\beta$. We see that by balancing the free energy of the system, the g's do not allow the system to fall in the lowest energy state and stay there forever. The first correction is of the form

$$C_V^\alpha (\delta\beta)^2 , \tag{41}$$

and keeping this term of order one implies that in the large volume limit $\delta\beta$ has to be kept of order $C_V^{-\frac{1}{2}}$, see also later.

Now the crucial observation: at each β_α value (after the initial thermalization) the system is always in equilibrium with respect to the *usual* Boltzmann distribution. The system is always (even at the moment of β-transitions) at thermal equilibrium.

The analysis of observable quantities is very easy. One just has to select all configurations which were flagged by a given β value:

$$\langle O \rangle_{\beta_\alpha} = \frac{\sum \text{All configurations at}_{\beta_\alpha} O_{\text{conf}}}{\text{Number of configurations at}_{\beta_\alpha}} . \tag{42}$$

How do we select a good set of $\{\beta_\alpha\}$ values? Fig. 7 can help to give an answer. Let the first $P(E)$ on the left be the one at the lowest T value (the one we are interested in thermalizing). The center one is at $T + \delta$, i.e. at the second T_α value. We require that the overlap of the two probability distributions be non-negligible. A configuration with an energy included in the colored region of the picture is typically a good configuration both at T and at $T + \delta$ (it is here that problems connected to discontinuities, like in first order phase transitions make the method fail). So, $\delta \equiv T_{\alpha+1} - T_\alpha$ has to be selected such that *there is a non-zero overlap between the two energy probability distributions* (as usual in Metropolis like methods one looks for a reasonable acceptance factor, of order 0.5, for the β_α moves). That should also make more clear why we might like to do more than one $\{\sigma\}$ sweep before updating β: we want to avoid a series of jumps between adjacent β_α values, and give the system the possibility of moving at fixed β to the opposite extremum of the $P(E)$ before trying changing β again.

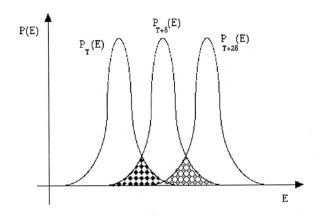

Fig. 7. $P(E)$ for different values of T.

We repeat the basic physical mechanism we are interested in. We are thinking about systems with very high free energy barriers. Warming the system up lowers the barriers, which disappear at T_c. When cooling down (always staying at thermal equilibrium!) the system can fall in a completely new valley.

Experimentally the method turns out to work very well for $3d$ Edwards-Anderson spin glasses. It is interesting to note that temperature chaos [36–38], should give troubles, but on the lattice sizes we have been able to study we did not get them: it is interesting to try and understand this issue better. In the case of $3d$ Edwards-Anderson spin glasses by using parallel tempering (see (3.3)) we were able to thermalize a 16^3 system down to $0.7T_c$ (Marinari et al. [39]). On a 24^3 lattice one is able to thermalize with a reasonable computer time (we are talking about single runs on one disordered sample taking the order of one day on a workstation) down to $0.9T_c$, while for lower T the situation gets more difficult to control.

The method does not work, [40], for generic heteropolymers (with order 30 sites, Lennard-Jones potential with quenched random couplings, [41]), and this is probably connected to the fact that the glassy transition is in this case of a discontinuous nature: for true (small) proteins, however, the method seems to be helpful, [42]. The method has also been applied by Fernandez et al. [17], by Caracciolo et al. [21] and by Coluzzi [18] to the $4d$ disordered Heisenberg model, by Kerler and Rehberg [20] to the $2d$ EA spin glass (by noticing one of the important advantages of the method, i.e. the trivial vectorization and parallelization of the scheme), and by Vicari [19] to CP^{N-1} models.

3.2 Random field Ising model

The case study we have done in our first tempering paper [16], discussed the *random field Ising model* which is ideal for illustrating in some more detail the main feature of the method. The model is defined by the Hamiltonian

$$H = -\sum_{\langle i,j \rangle} \sigma_i \sigma_j + \sum_i h_i \sigma_i , \qquad (43)$$

where $\sigma_i = \pm 1$, the first sum runs over nearest neighbor sites on a simple cubic lattice in d dimensions, and the local external fields h_i are quenched random variables, which take the value $\pm|h|$ with equal probability.

We have taken $d = 3$, $V = 10^3$ sites and $|h| = 1$, sitting close to the critical region, on the low temperature side (we will focus here on studying $\beta = .26$). Fig. 8 gives an idea of the critical region. The specific heat has a maximum close to $\beta = .25$. In these first runs in most cases we allowed the system to visit only 3 β values, $(\beta_\alpha = (.24, .25, .26))$, and sometimes 5 β values: for the Edwards Anderson model in the most recent runs on a 24^3 lattice we use up to 50 β values [39]. The 3 β values we have given above are selected such as to span the range from the low temperature to the high temperature phase. This is the general principle: the system has to be allowed to travel, in β space, from the cold phase, that has a physical interest, to the warm phase, where free energy barriers disappear and correlation times are short. In Fig. 9 we plot the β_α values the system selects in the course of the dynamics. Notice that the system is not getting trapped in the cold or in the warm phase: acceptance of the β swap is easy, and the system easily travels between the two phases.

The magnetization measured from typical normal, fixed-β Monte Carlo Metropolis runs is shown in Fig. 10: here the correlation time, since one never sees a flip to the opposite magnetization, seems short. That taking Fig. 10 at face values would be misleading is clear from the magnetizations from a tempering run shown in Fig. 11: here tempering allows the system to flip between the $\pm m$ states, and it is clear that configurations with positive m have a nonnegligible weight in the equilibrium distribution.

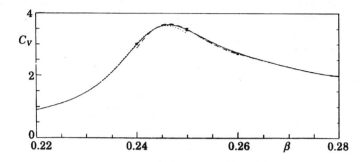

Fig. 8. The measured specific heat C_V as a function of β. $\beta = 0.26$, in the low temperature region, is the point on which we focus.

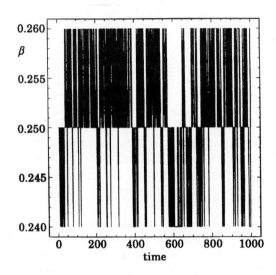

Fig. 9. β_α as a function of the Monte Carlo tempering time.

In Fig. 12 we plot the magnetization of the configurations that had a β value of 0.26. Here one clearly sees that the state with positive m is allowed at $\beta = 0.26$ with a smaller but definitely non-zero probability than for the negative m state. These last figures give the main idea. In a non-disordered Ising model flipping from the minus state to the plus state would be irrelevant, if we are not interested in studying details of the tunneling dynamics: we know by symmetry that to the minus state corresponds an identical plus state. But this is not true for a disordered model, where the tunnelling between degenerate ground states that are not related by a trivial symmetry is the most interesting part of the

Optimized Monte Carlo Methods 69

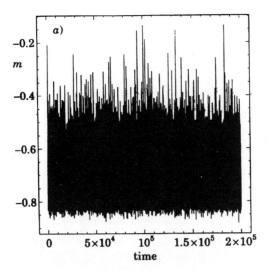

Fig. 10. Magnetization as a function of Monte Carlo time, at $\beta = 0.26$, for a normal Metropolis algorithm. The system never tunnels to a positive magnetization value.

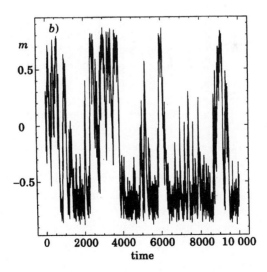

Fig. 11. Magnetization versus Monte Carlo time for tempering. Here all β values are included: the measured magnetization are related to configurations that are at thermal equilibrium with different temperature values.

dynamics. Here the random quenched magnetic field breaks the symmetry, and we want to explore the different ground states. Tempering allows us to do that.

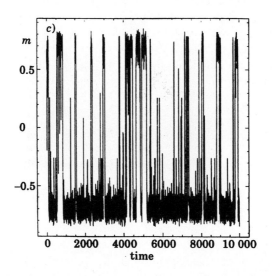

Fig. 12. As in Fig. 11, but we have selected only configurations with $\beta = 0.26$.

This was a very easy numerical experiment, but the steps we described are also relevant for more complex situations: for example we said earlier that the numerical simulations we are running now ([39]) on large lattices for the $3d$ Edwards Anderson spin glass are quite complex (but they work very well).

Tempering is related to annealing. A trivial extension of annealing to non-zero values of T is impossible: annealing only gives information about energy, while for $T > 0$ we want to deal with the free energy. We do not get from annealing any information about the entropic structure of phase space. Tempering can be seen as such a generalization.

Tempering could also turn out to be an effective global optimization scheme, even if this issue has not been looked at in much detail so far. For some very preliminary unpublished studies conult [43], [44]. The most important point is that tempering contains a built-in, self-implemented schedule. Setting up the schedule is the most serious problem with annealing, and tempering does it for us.

3.3 Parallel tempering

Parallel tempering has been discussed in [22], [23], [24], [25]. We will describe it here. As we have already said, it is so simple that it makes many of the details we have discussed with simple tempering of no practical importance. Also, parallel tempering performs dramatically well, for example, on finite dimensional Edwards Anderson models.

In figs. (6) and (7) we have shown how we select the T values we use during the simulation. Let us say that by using the criteria we have defined before we need for example $N_{(\beta)}$ values of β_α. Now in the parallel tempering approach, you run $N_{(\beta)}$ simulations in parallel ($N_{(\beta)}$ different configurations C_α of the system that evolve in the same quenched disordered landscape). Each copy starts with a different β value assigned to it. For example start with

$$\beta(C_0) = \beta_0;\ \beta(C_1) = \beta_1;\ \ldots\ \beta(C_\alpha) = \beta_\alpha;\ \ldots\ \beta(C_{N_{(\beta)}}) = \beta_{N_{(\beta)}}\ . \qquad (44)$$

Now the composite Monte Carlo method is based on two series of steps:

1. Update the $N_{(\beta)}$ copies of the system with a usual Metropolis sweep of

$$\{\sigma\}_{C_\alpha} \ldots \{\sigma\}_{C_{N_{(\beta)}}}\ . \qquad (45)$$

2. Swap the β values.
   - Propose the first β-swap: the configuration that is at β_0 can go to β_1 and the one that is at β_1 can go to β_0. Each spin configuration carries a β-label. Configurations carrying adjacent β-labels try to swap their labels. You use Metropolis for swapping β with the correct probability (see later).
   - Propose the second β-swap: the configuration that is at β_1 can go to β_2 and the one at β_2 can go to β_1.
   - And so on, to the last couple of configurations: the configuration that is at $\beta_{N_{(\beta)}}$ can go to $\beta_{N_{(\beta)}-1}$ and the one at $\beta_{N_{(\beta)}-1}$ can go to $\beta_{N_{(\beta)}}$.

I stress the fact that you always try to swap *adjacent* β-values (otherwise the procedure would be not effective). At a given point in time configuration number 27 can be at β_0 and configuration number 3 at β_1: these are the two configurations you will try to swap.

The β-Metropolis swap: as usual, you will accept if the swap makes the energy decrease, and will accept with probability $e^{-\Delta S}$ if it makes the energy increase. You will have to compute

$$\begin{aligned}\Delta S &= S' - S \\ &= (\beta_{\alpha+1} E_{C_{\beta_\alpha}} + \beta_\alpha E_{C_{\beta_{\alpha+1}}}) - (\beta_\alpha E_{C_{\beta_\alpha}} + \beta_{\alpha+1} E_{C_{\beta_{\alpha+1}}})\ ,\end{aligned} \qquad (46)$$

where $E_{C_{\beta_\alpha}}$ and $E_{C_{\beta_{\alpha+1}}}$ obviously do not change during the β-swap.

Here there is no freedom (and no need) for an additive term like the one we had before, g_α. In this method once the β values have been fixed (at that can be done loosely, without the need for fine tuning) there are no free parameters, and no fine tuning needed. The ensemble of parallel tempering is very different from the one of tempering, even if the two methods look very similar.

A possible way to look at the fact that we do not need the g parameters is the following. The g_α were needed in order to prevent the system from collapsing to the state with lowest energy. But here there is a fixed number of β-values. A given value of β_α, for example β_{23}, cannot disappear. This fact makes things easy for us.

3.4 Thermalization

Thermalization is a crucial issue. This is true even for usual spin models and usual local dynamics. We want to be sure that we are looking at a system that has reached equilibrium, and after that we want to be sure that we have correlation times under control. In other words we are interested in studying the asymptotic equilibrium probability distribution, and we need to be sure that we are collecting the right information. In complex dynamics such as tempering, involving multiple Markov chains, it is even more difficult to keep correlation times under control. The fact that we are typically studying complex systems, where slow dynamics is one of the most prominent features, does not help. I will discuss here some details of the understanding we have reached about this issue. We will try to understand which kind of thermalization criteria we can adopt when using tempering.

Let us start by describing what happens when we use simple Metropolis dynamics for studying a system with quenched disorder. Here we will not give details, they can be found for example in Mézard et al. [45], but I will only remind the reader that in this case one mainly deals with functions of the *overlap*

$$q^{\alpha\beta} \equiv \frac{1}{V} \sum_i \left(\sigma_i^\alpha \sigma_i^\beta \right) , \tag{47}$$

where α and β label two replicas of the systems, i.e. two equilibrium configurations defined under the same realization of the quenched disorder. The overlap gives us information about how similar two typical equilibrium configurations of the same system are. q plays the role of the order parameter, and it is the equivalent of the magnetization m for usual spin models. The typical shape of the probability distribution of q, averaged over the quenched disorder J (see [45]), is shown in Fig. 13. The two peaks would be sharp (at $\pm m^2$) for a usual spin model. The non-trivial part here is the fact that there is a continuous, non-zero contribution close to $q \simeq 0$: the system can exist at equilibrium in many states, that can even be, in the infinite volume limit, completely different.

In the case of a normal Metropolis update one can use two very strong criteria to check thermalization.

1. *Check symmetry of $P(q)$ under $q \leftrightarrow -q$ for each sample.* This is a very strong criterion. The main issue here is that the full flip of the whole system, $\{\sigma\} \to \{-\sigma\}$ is the slowest mode of the dynamics. If you have done that well enough to get a good $\pm\sigma$ symmetry you have explored the whole phase space. In the simulation of a normal spin-model in the cold phase we would never

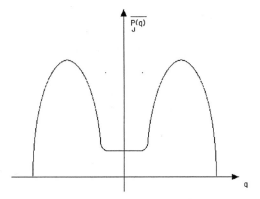

Fig. 13. $\overline{P_J(q)}$ versus q. The flat part close to $q = 0$ is the most remarkable feature of the behavior of a class of disordered systems.

be so demanding (if not interested in details of tunneling amplitudes): we would just sit in one vacuum and compute observable quantities there. As we have already said, this is the main difficulty connected to the study of disordered systems.

In Fig. 14 we give two typical $P(q)$ for two different given quenched realizations of the disorder. In the Parisi solution of the mean field model (see [45]) you can compute many properties in detail: for example how many configurations contribute to the $q \simeq 0$ plateau. The numerical results in $3d$ show a very good similarity with the mean field results ([46], [39]).

This criterion can be adopted for tempering, but in this case it is not so strong anymore: in tempering methods the mode $\{\sigma\} \rightarrow \{-\sigma\}$ is not necessarily a very slow mode. It is possible in this case that one gets a very symmetric $P(q)$ that is not related to the asymptotic equilibrium distribution function.

2. A second criterion, originally due to [47], that is very strong for usual dynamics, is based on using two different definitions of the $P(q)$.

   (a) After a random start from two different spin configurations simulate two independent copies of the system (in the same set of $\{J\}$ couplings). This is the definition we had in mind till now. If we denote by σ and τ the two copies of the system we call q_2

   $$q_2 \equiv \frac{1}{V} \sum_i (\sigma_i \tau_i) , \tag{48}$$

   and $P_2(q)$ the equilibrium probability distribution of q_2.

   (b) In the second approach we use a dynamical measurement, at different times. We wait for a large Monte Carlo time separation t and we define

(for t_0 large enough)

$$q_{dyn} \equiv \frac{1}{V} \sum_i \langle \sigma_i(t_0)\sigma_i(t_0+t)\rangle , \qquad (49)$$

where eventually we will take an average over t_0. We define $P_{dyn}(q)$ the equilibrium probability distribution of q_{dyn}.

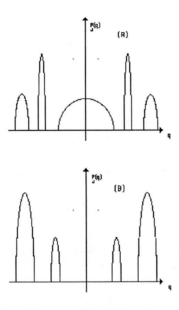

Fig. 14. Two typical $P_J(q)$ versus q, for two different realizations of the couplings.

When using the first definition at short times the two copies are not as similar as they will be asymptotically. Correlations only build up slowly. So $P_2(q) \to P(q)$ from below, and at short times $P_2(q)$ is centered at lower q values than $P(q)$. On the other hand, at short times $\sigma(t_0)$ and $\sigma(t_0+t)$ are correlated, i.e. q_{dyn} at short times tends to be larger than the asymptotic value. So $P_{dyn}(q) \to P(q)$ from above. Only for times large enough is $P_{dyn}(q) = P_2(q)$, and we use this condition to check thermalization.

Unfortunately the extension of this second criterion to tempering is not straightforward, since now dynamics is assigning different β values to different configurations.

For parallel tempering runs, we use three main conditions to check thermalization.

1. We check that no observable is drifting in time. We look for example at the energy E and q^2. We check, on logarithmic time scales, that they have safely

converged to a stable value. We also explicitly look at the full $P_J(q)$, and check it has no drift.
2. We check that the spacing of the allowed β_α's is small enough to guarantee a good acceptance ratio for the proposed β swaps.
3. We demand that each of the $N_{(\beta)}$ configurations C_α must visit each of the allowed β_α values with similar frequencies. If we have done ten million sweeps and we have ten allowed β values we demand that each configuration has spend more or less one million sweeps in each β value. We keep a set of counters to check that. We want to be able to detect situations where some systems are confined in a part of the phase space, and a good acceptance factor is not enough to avoid that (the configurations could just be flipping locally in β space).

3.5 More comments

It is interesting to make some other technical comments about tempering like methods, mainly having in mind the large volume scaling behavior and the performances of the method. Here we go, with a slightly miscellaneous series of comments.

When you increase the lattice size you need a larger number of allowed β_α values, i.e. a larger $N_{(\beta)}$, to sample the β-space, and to reach the region where free energy barriers disappear. The method is critically slowed down, but only like a power law.

Again, disordered systems need more attention than normal systems. You have to check that for each realization of the quenched disorder $\{J\}$ the condition that all C_α have visited all β_α values for similar periods of time is satisfied. From our experience the situation seems quite sharp: when the method works it works well, when it does not work it is a disaster (i.e. some visiting times are of order zero and some are of the order of the total time). For normal tempering (but parallel tempering seems preferable in all known cases) the constants $\{g_\alpha\}$ have to be tuned separately in each sample.

Let us discuss in more detail the volume scaling in tempering-like methods. Let us select the allowed β values at a constant distance δ:

$$\beta_{\alpha+1} = \beta_\alpha + \delta \ . \tag{50}$$

The probability for accepting a β-swap is

$$P_{SWAP}(\beta_\alpha, \beta_{\alpha+1}) \equiv e^{-\Delta} \ , \tag{51}$$

and

$$-\log(P_{SWAP}) = \Delta = \delta \cdot \left(S(C_{\beta_{\alpha+1}}) - S(C_{\beta_\alpha})\right) \simeq \delta^2 \frac{dE}{d\beta} \ . \tag{52}$$

So if the specific heat does not diverge we want to select

$$\delta^2 N = \text{constant} \ , \ \delta \simeq N^{-\frac{1}{2}} \ , \ \text{i.e. } N_\beta \simeq N^{\frac{1}{2}} \ . \tag{53}$$

At a second order phase transition point, where the specific heat diverges, we have to be slightly more careful, since the number of intervals we need turns out to be higher. One has

$$\xi \simeq |T - T_c|^{-\nu} , \quad C_V \simeq |T - T_c|^{-\alpha} ,$$
$$|T - T_c| \simeq \xi^{-\frac{1}{\nu}} \simeq N^{\frac{1}{d\nu}} ,$$
$$C_V \simeq N^{\frac{\alpha}{d\nu}} . \tag{54}$$

So we get

$$\delta^2 N^{1+\frac{\alpha}{d\nu}} = \text{constant} , \quad N_\beta \simeq \delta^{-1} \simeq N^{\frac{1}{2}(1+\frac{\alpha}{d\nu})} , \tag{55}$$

that is our final estimate.

The choice of the set $\{\beta_\alpha\}$ is not crucial (not like the g_α's in the serial tempering). We can select the same set for all the realizations of the disorder, paying only the price of loosing some small amount of efficiency.

In Fig. 15 we sketch the correlation times computed by Hukushima and Nemoto [25]. Filled dots are for tempering, empty dots for multicanonical. Simulations are for a $3d$ Edwards Anderson spin glass, with couplings $J = \pm 1$. 32 β values have been allowed in the parallel tempering run, for all N values. τ is defined as the typical time needed from a spin configuration for going from the cold to the warm phase. It is worth noticing that in parallel tempering, given the β set, there are no parameters to be tuned.

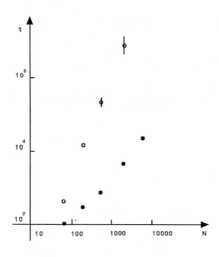

Fig. 15. Correlation times for multicanonical simulation and for tempering, for different lattice sizes (on a log-log plot). The figure is only sketchy, and the interested reader can find the exact data points in Fig. 2 of the original paper [25]. Filled dots are for tempering, empty dots for multicanonical.

3.6 Umbrella sampling and reweighting

All the ideas we have discussed so far are related to the technique of umbrella sampling by Torrie and Valleau [27], [28] (mainly developed for simulations of liquid systems). Let us show now how we can relate that to reweighting techniques.

We consider an observable quantity A

$$\langle A \rangle_\beta = \frac{\int \{dC\} e^{-\beta S(\{C\})} A(\{C\})}{\int \{dC\} e^{-\beta S(\{C\})}}$$
$$= \int \{dC\} \pi_\beta(\{C\}) A(\{C\}) , \qquad (56)$$

where π_β is the Boltzmann distribution at β,

$$\pi_\beta = \frac{e^{-\beta S(\{C\})}}{Z_\beta} . \qquad (57)$$

So, now, we want to improve things: maybe dynamics at β are slow, or we already have data at $\tilde{\beta}$ and we want an added bonus, or maybe we want to measure something fancier (see earlier in the text). So we write

$$\langle A \rangle_\beta = \frac{\int \{dC\} e^{-\beta S(\{C\})} A(\{C\}) \frac{\tilde{\pi}}{\tilde{\pi}}}{\int \{dC\} e^{-\beta S(\{C\})} \frac{\tilde{\pi}}{\tilde{\pi}}} \left(\frac{\int \{dC\} \tilde{\pi}}{\int \{dC\} \tilde{\pi}} \right) , \qquad (58)$$

where we have inserted a number of ones. We find in this way that

$$\langle A \rangle_\beta = \frac{\langle \frac{A e^{-\beta S}}{\tilde{\pi}} \rangle_{\tilde{\pi}}}{\langle \frac{e^{-\beta S}}{\tilde{\pi}} \rangle_{\tilde{\pi}}} , \qquad (59)$$

where the expectation values are now taken over the $\tilde{\pi}$ probability distribution. If we choose

$$\tilde{\pi} = \frac{e^{-\tilde{\beta} S}}{Z_{\tilde{\beta}}} \qquad (60)$$

we find the usual reweighting, as we have already discussed. But now we can also use the most general *umbrella sampling* by [27], [28]. The relation we have just derived, (60), is valid for a generic probability distribution $\tilde{\pi}$. $\tilde{\pi}$ does not need to have the form of a Boltzmann distribution at some value of β, it can be anything you like. This is umbrella sampling: open $\pi \to \tilde{\pi}$, as an umbrella, to cover all of the parameter space in the region of interest. For example you can take

$$\tilde{\pi}(\{C\}) = \sum_\alpha w(\beta_\alpha) e^{-\beta_\alpha S(\{C\})} . \qquad (61)$$

Selecting the w such that you get an equal sampling of the β points gives $w \simeq e^f$, i.e. the choice of tempering.

3.7 Multicanonical methods

A large amount of work has been done on the so called *multicanonical methods*, [31–34], that are very powerful in studying discontinuous phase transitions. We discuss here a simple example where the $2d$, 10 states Potts model is analyzed. The results we describe, due to Berg and Neuhaus [32], are for lattices up to 100^2. The action has the form

$$S = \sum_{\langle i,j \rangle} \delta_{s_i,s_j}, \quad s_i = 0, 1, \ldots, 9 \ . \tag{62}$$

Such a model undergoes a temperature driven strong first order phase transition. Computing the interfacial free energy between the disordered and the ten ordered states is a hard problem.

On a finite lattice there are no phase transitions: we define β_L^c, the pseudocritical coupling on a lattice of linear size L, such that the two peaks in the probability distribution of the internal energy have the same height. In Fig. 16 (taken from [32]) we show the probability distribution of the energy for different lattice sizes: on larger lattices the probability of getting a configuration in the forbidden region becomes smaller and smaller. One has that

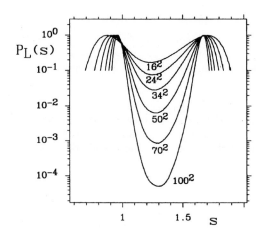

Fig. 16. The probability distribution of the energy for different lattice sizes.

$$P_L^{(min)} \simeq e^{-\sigma L^{d-1}} \ . \tag{63}$$

When using normal Metropolis dynamics (63) implies that the tunneling time diverges severely with the lattice size:

$$\tau_{Metropolis} \simeq A L^a e^{\sigma L^{d-1}} \ . \tag{64}$$

The multicanonical method takes a different approach, by modifying the sampling probability distribution. Here one samples phase space with weight

$$P_L^{MCan} \simeq e^{a_L^k - \beta_L^k S}, \text{ for } S_L^k < S \leq S_L^{k+1} \tag{65}$$

(instead of the usual $P_L^{Bolt} \simeq e^{-\beta_L S}$). One has partitioned the action range in intervals I_k, using a different action in each range. Now one chooses intervals I_k and parameters a_L^k, b_L^k such that $P_l^{(MCan)}$ is flat: Fig. 17 shows that it can be done very successfully.

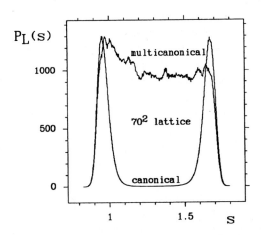

Fig. 17. The Boltzmann and the multicanonical probability distribution of the energy for $L = 70$. The multicanonical distribution is quite flat, allowing easy transitions between the two sides of the first order phase transition.

Configurations that were exponentially suppressed are now enhanced by the multicanonical action: the interested reader can look for the details of fixing the parameters in the original papers, [31–34].

In the case of the multicanonical algorithm after data collection you need reweighting (as opposed to tempering, where the output data at each β value are thermally distributed) to reconstruct the Boltzmann distribution. At β_L^c

$$P_L^{Bolt} = e^{\beta_L^c S - \beta_L^k - a_L^k} P_L^{MCan}. \tag{66}$$

The improvement is dramatic, and the exponential slowing down becomes power-like ([32]; the same happens for tempering, for example in the plus to minus tunneling in the low temperature phase in the Ising model, [17]). An estimate of Berg and Neuhaus [32] gives $\tau_L^{MCan} \simeq 0.7 L^{2.7}$, versus $\tau_L^{HeatBath} \simeq 1.5 L^{2.15} L^{0.08L}$.

References

[1] Falcioni, M., Marinari, E., Paciello, M. L., Parisi, G., Taglienti, B., *Phys. Lett.* **108B** 331 (1982).
[2] Marinari, E., *Nucl. Phys.* **B 235** 123 (1984).
[3] Salsburg, Z., Jackson, J. D., Fickett, W., Wood, W. W., *J. Chem. Phys.* **30** 65 (1959).
[4] Chestnut, D., Salsburg, Z., *J. Chem. Phys.* **38** 2861 (1963).
[5] Mc Donald, I., Singer, K., *J. Chem. Phys.* **47** 4766 (1967).
[6] Ferrenberg, A., Swendsen, R., *Phys. Rev. Lett.* **61** 2635 (1988).
[7] Ferrenberg, A., Swendsen, R., *Phys. Rev. Lett.* **63** 1195 (1989).
[8] Itzykson, C., Pearson, R. B., Zuber, J.-B., *Nucl. Phys.* **B 220** 415 (1983).
[9] Pearson, R. B., *Phys. Rev.* **B 26** 6285 (1982).
[10] Le Guillou, J. C., Zinn-Justin, *Phys. Rev.* **B 21** 3976 (1980).
[11] Bhanot, G., et al., *Phys. Rev. Lett.* **59** 803 (1987).
[12] Rothe, H. J., *Lattice Gauge Theory: An Introduction.* (World Scientific, Singapore, 1992).
[13] Montvay, I., Münster, G., *Quantum Fields on a Lattice.* (Cambridge University Press, Cambridge UK, 1994).
[14] Kogut, J., *Rev. Mod. Phys.* **51**, 659 (1979).
[15] Parisi, G., *A Short Introduction to Numerical Simulations of Lattice Gauge Theories.* In *Critical Phenomena, Random Systems, Gauge Theories*, edited by K. Osterwalder and R. Stora, proceedings of Les Houches, Session XLIII, 1984 (Elsevier, Amsterdam, 1986).
[16] Marinari, E., Parisi, G., *Europhys. Lett.* **19** 451 (1992).
[17] Fernandez, L. A., Marinari, E., Ruiz-Lorenzo, J., *J. Phys. I (France)* **5** 1247 (1995).
[18] Coluzzi, B., *J. Phys. A (Math. Gen.)* **28** 747 (1995).
[19] Vicari, E., *Phys. Lett.* **B 309** 139 (1993).
[20] Kerler, W., Rehberg P., *Simulated Tempering Approach to Spin Glass Simulations.* (cond-mat/9402049).
[21] Caracciolo, S., Pelissetto, A., Sokal, A. D. (1994). Unpublished note.
[22] Tesi, M. C., Janse van Rensburg, E. J., E. Orlandini E., Whillington, S. G., *Monte Carlo Study of the Interacting Self-Avoiding Walk Model in Three Dimensions.* Oxford preprint OUTP-95-06S, *J. Stat. Phys.* (1995).
[23] Geyer, J., Thompson, E. A., University of Minnesota preprint, 1994.
[24] Hukushima, K., Takayama H., Nemoto, K., *Application of an Extended Ensemble Method to Spin Glasses. Int. J. Mod. Phys.* **C** (1995).
[25] Hukushima, K., Nemoto, K., *Exchange Monte Carlo Method and Application to Spin Glass Simulations.* (cond-mat/9512035).
[26] Krauth, W., (1996). Contribution to this volume.
[27] Torrie, G. M., Valleau, J. P., *J. Comp. Phys.* **23** 187 (1977).
[28] Torrie, G. M., Valleau, J. P., *Chem. Phys.* **66** 1402 (1977).
[29] Graham, J. P., Valleau, J. P., *Chem. Phys.* **94** 7894 (1990).

[30] Valleau, J. P., In *Proceedings of the International Symposium on Ludwig Boltzmann*, edited by G. Battimelli, M. G. Ianniello and O. Kreiten, (1993).
[31] Berg, B. A., Neuhaus, T., *Phys. Lett.* **B 267** 249 (1991).
[32] Berg, B. A., Neuhaus, T., *Phys. Rev. Lett.* **68** 9 (1992).
[33] Berg B. A., Celik, T., *Rev. Lett.***69** 2292 (1992)
[34] Berg B. A., Celik, T., Hansmann, U., *Europhys. Lett.* **22** 63 (1993).
[35] Kerler, W., Weber, A., *Phys. Rev.* **B47** 11563 (1993).
[36] Kondor, I., *J. Phys.* **A22** L163 (1989).
[37] Kondor, I., Végsö, A., *J. Phys.* **A26** L641 (1993).
[38] Ritort, F., *Chaos in Short Range Spin Glasses.* (cond-mat/9307065).
[39] Marinari, E., Parisi, G., Ruiz-Lorenzo, J., (1996). In preparation.
[40] Iori, G., Marinari, E., Parisi, G., (1996). In preparation.
[41] Iori, G., Marinari, E., Parisi, G., *J. Phys.* **A24** 5349 (1991).
[42] Hansmann, U. H. E., Okamoto, Y., *J. Comp. Chem.* **14** 1333 (1993).
[43] Rose, T., Coddington, P. D., Marinari, E., *Evaluation of Simulated Tempering for Optimization Problems.* NPAC preprint (Syracuse, NY, USA, 1992), unpublished, available at:
ftp://ftp.npac.syr.edu/pub/projects/reu/reu92/papers/rose.ps.
[44] Shore, M., Coddington, P. D., Fox, G. C., Marinari, E., *A New Automatic Simulated Annealing-Type Global Optimization Scheme.* NPAC preprint (Syracuse, NY, USA, 1993), unpublished, available at:
ftp://ftp.npac.syr.edu/pub/projects/reu/reu93/papers/Shore.ps.Z.
[45] Mézard, M., Parisi, G., Virasoro, M. A., *Spin Glass Theory and Beyond.* (World Scientific, Singapore, 1987).
[46] Marinari, E., Parisi, G., Ritort, F., Ruiz-Lorenzo, J., *Phys. Rev. Lett.* **76** 843 (1995).
[47] Bhatt, R. N., Young, A. P., *J. Magn. Matter* **54** 191 (1986).

Monte Carlo on Parallel and Vector Computers

Dietrich Stauffer

Institute for Theoretical Physics, Cologne University, D-50923 Köln, Germany

Two hardware developments have speeded up computing during the last 20 years apart from making all circuits faster: vector and parallel computing. We summarize here the essentials; more details are given in our textbook. Actually most vector computers today are also parallel computers, but for the purposes of teaching and programming these two aspects can be separated here.

Vector computing can gain an order of magnitude in speed by treating long arrays of data like producing cars on an assembly line. If $A_i = B_i + C_i$ is calculated for a long loop over i, then a normal computer calculates the addresses of A_i, B_i, C_i, fetches B_i and C_i from memory, adds both numbers, and stores the sum in memory. All that happens in separate machine cycles before i can be incremented. A vector computer deals simultaneously with many of these operations by fetching e.g B_{104} and C_{103}, adding $A_{102} = B_{102} + C_{102}$, and storing A_{101} in memory, all at the same time. Similarly to a factory's assembly line, such vector computing is useful only for a large number ($> 10^2$) of elements on which the same operations need to be performed. It fails if the computation of step i requires the results obtained in step $i-1$.

For example, the diffusion of a single particle through a random medium cannot be simulated vectorially since each position depends on the previous position. However, if an outer loop runs over many different independent particles to be averaged over, then this outer loop can be exchanged with the inner loop over time, and now (with additional memory requirements) we can vectorize efficiently over all particles, instead of over time. This exchange of outer and inner loops is often helpful, and sometimes done automatically by the compiler. If lattice models of interacting elements have to be added sequentially (like in Ising models) and not simultaneously (like in cellular automata), then it may be necessary to divide the system into sublattices such that no particles on one sublattice interact directly with each other.

IF-conditions slow down the flow of operations, except if they depend on a Fortran "parameter" and thus are already decided at compile time. Input/output operations, user-written subroutines/functions, and jump commands destroy vectorization. In general, many problems of statistical physics cannot be vectorized efficiently. This is different for parallel computing, based on the MIMD principle (multiple instruction, multiple data) with distributed memory (each processor has its own storage). The *replication* method simulates N lattices on N different processors of the parallel computer using the same program but e.g. different random number seeds. Global summation routines like gisum on the

Intel Paragon allow easy adding up of equivalent numbers on all N different processors.

More complicated is the simulation of one very large system distributed onto the N different processors by *domain decomposition* (geometric parallelization). Then these processors have to communicate their status to their neighbor processors since particles on different processors may interact with each other. This communication is easiest if we have interactions only between nearest neighbors on a lattice. Then a lattice of linear dimension L is divided into N strips or plates of width $Lstrip = L/N$ each, with each processor working on one strip only. The processors are imagined as being arranged from top to bottom. The processor in the middle needs a top buffer and a bottom buffer. The top buffer contains the lowermost line (or plane) of the upper neighbor processor, and the lower buffer stores the uppermost line/plane of the lower neighbor processor. When a processor has finished the evaluation of its first line/plane, it sends it via message passing to the lower buffer of its upper neighbor; analogously the last calculated line/plane is sent to the upper buffer of the lower neighbor processor. These message passing commands urgently need standardization.

The parallel efficiency is the calculation speed of N processors together, divided by N times the speed of one processor. Users should try to keep it above 0.5. In special cases efficiencies above one can be found (superlinear speedup).

Important deatils on supercomputer Monte-Carlo methods can be found in: Stauffer, D., Hehl, F. W., Ito, N., Winkelmann, V., Zabolitzky, J. G., *Computer Simulation and Computer Algebra*, 3rd edition (Springer, Berlin, Heidelberg, 1993).

Appendix

A News

What can be done with high-performance computers in statistical physics? Presumably the most studied standard problem is the 76 year old Ising model where each lattice site carries a spin ± 1; the magnetization is the number of up spins minus the number of down spins. In two dimensions we do not need computers to find the critical temperature below which a nonzero spontaneous magnetization appears. In three dimensions (simple cubic lattice with exchange energy J between nearest neighbors) Talapov and Blöte [1] give $J/kT_c = 0.2216544$ more accurately than series expansions. In four and five dimensions the Cray-T3E is competitive with series expansions to find T_c [2] while in six and more dimensions series expansions still win. The dynamic critical exponent z is about 2.18 and 2.05 in two and three dimensions, from lattices of 496640^2 and 5888^3 spins [2]; even $10^6 \times 10^6$ was simulated using an entirely different method [3].

How does the magnetization $M(t)$ relax towards its equilibrium value if we start with all spins up? Above T_c we observe a simple exponential without a power-law prefactor, for all dimensions. Below T_c we confirmed a simple exponential in four and more dimensions, a stretched exponential $\propto \exp(-\text{const} \cdot \sqrt{t})$

in two dimensions, and a simple exponential multiplied with a power law in three dimensions [2]. These shapes agree nicely with droplet model predictions and other theories.

In finite systems, one has to distinguish between the relaxation of the magnetization and of its absolute value, and both are within the error bars of the computer simulation in agreement with renormalized ϕ^4 field theory [4].

All of the above work refers to "model A" where the magnetization and the energy fluctuate, for a fixed field and fixed temperature. Cellular automata simulations for fixed energy are in progress [5].

Another well-studied problem is percolation [6], where each site of a lattice is randomly occupied or empty with probabilities p and $1-p$, respectively. For $p < p_c$ only finite clusters exist while for $p > p_c$ one infinite cluster also is formed. Here a cluster is a group of neighboring occupied sites. $R(p)$ is the probability of having a cluster spanning from top to bottom. Recent years have shown two widespread opinions to be wrong: First, site-to-cell renormalization in the sense of the fixed point equation $R(p_c) = p_c$ does not become correct for large lattices [7]. Second, in contrast to the impression left by [8], there can be more than one spanning cluster in two to five dimensions [9] right at $p = p_c$.

B A program

The enclosed multispin-coding program calculates the spreading of an infection on a $L \times L$ square lattice with free boundary conditions: A site becomes occupied if at least one of its four neighbors is occupied. Since 32 sites are stored in one word $m(i,j)$, the index i runs from 1 to $LL = L/32$ only, while j goes from 1 to $Lstrip = L/np$. Loop 7 is the core of the program, where the new computer word is the bit-by-bit OR of the four neighbors at $i\pm 1$, $j\pm 1$. Before loop 6 we have the two send and the two corresponding receive commands for the buffer lines, in the Intel Hypercube/Paragon language: The first argument identifies the message, the second one gives the first element of the array to be sent or received, the third one the number of bytes to be transmitted, the fourth one the processor number to which the information is sent. The function mynode() =0, 1, ... np-1 gives the number of the processor one is working on, whereas numnodes() = np indicates how many processors are used by this program. dclock() gives the time. (For examples of multi-spin coding without parallelization see *J. Phys.* **A 24** 909 (1991).)

In this way, 125 Intel 860 processors infected in two minutes (plus initialization time) all sites of a 100,000 * 100,000 lattice even though initially only one percent were infected.

```
c     cellular automata on square lattice: infection by OR rule
c     L*L system distributed on np processor with L=32*LL and 32 bits/integer
      parameter(np=125,LL=3125,LL1=LL+1,L=LL*32,Lstrip=L/np,Lp=Lstrip+1)
      dimension n(0:LL1,0:Lp),m(LL,Lstrip)
      real*8 dclock,p
      common, allocatable /giant/ n,m
      data p/0.01d0/, max/100/, iseed/1/, mag/1/, mask/'FFFFFFFF'x/
      time=dclock()
      node=mynode()
      if(node.eq.0) print *, L, p, max, iseed, np
      if(np.ne.numnodes() .or. (L/np)*np.ne.L) stop 9
      allocate(/giant/)
      iprob=(2.0d0*p-1.0d0)*2147483648.0d0
      ibm=2*iseed-1
      do 5 i=0,node
5        ibm=ibm*65539
      length=4*LL
      do 1 j=1,Lstrip
      do 1 i=1,LL
1        n(i,j)=0
      do 2 nbit=1,32
      do 2 j=1,Lstrip
      do 2 i=1,LL
      n(i,j)=ishft(n(i,j),1)
      ibm=ibm*16807
c     random initialization of lattice: half one, half zero
2     if(ibm.lt.iprob) n(i,j)=n(i,j)+1
      if(node.eq.0) print *, dclock()-time
      do 3 itime=1,max
      mag=0
      do 4 j=1,Lstrip
      do 4 i=1,LL
c        mag counts the number of words still containing some zero bits
4     if(n(i,j).ne.mask) mag=mag+1
      call gisum(mag,1,idummy)
      if(node.eq.0) print *, itime,mag
      if(mag.eq.0) then
        if(node.eq.0) print *, dclock()-time, np
        stop
      end if
      if(node.lt.np-1) call csend(1,n(1,Lstrip),length,node+1,0)
      if(node.gt.0   ) call crecv(1,n(1,  0 ),length)
c     now message passing to and from upper and lower processor
      if(node.gt.0   ) call csend(2,n(1,  1 ),length,node-1,0)
      if(node.lt.np-1) call crecv(2,n(1, Lp ),length)
      do 6 j=1,Lstrip
c     free boundaries left and right via shift
      n(0,j)=ishft(n(LL,j),-1)
6     n(LL1,j)=ishft(n(1,j),1)
```

```
      do 7 j=1,Lstrip
      do 7 i=1,LL
c     finally, we use the OR rule to infect all neighbors of infected sites
7     m(i,j)=ior(ior(n(i,j-1),n(i,j+1)),ior(n(i-1,j),n(i+1,j)))
      do 8 j=1,Lstrip
      do 8 i=1,LL
8     n(i,j)=m(i,j)
3     continue
      deallocate(/giant/)
      stop
      end
```

```
     100000  1.0E-002       100         1       125   35.87135289999424
         1   312500000
         2   312500000
         3   312500000
         4   312500000
         5   312500000
         6   312500000
         7   312500000
         8   312500000
         9   312499998
        10   312499856
        11   312495902
        12   312440649
        13   312013770
        14   309950657
        15   303298089
        16   287936874
        17   261087777
        18   223742467
        19   180580541
        20   137565716
        21    99470696
        22    68709531
        23    45620720
        24    29272915
        25    18234487
        26    11067739
        27     6558578
        28     3797568
        29     2157897
        30     1202282
        31      658057
        32      355305
        33      189602
        34      100908
        35       52966
        36       27369
```

```
       37       14157
       38        7496
       39        4064
       40        2181
       41        1259
       42         722
       43         385
       44         254
       45         165
       46          99
       47          53
       48          23
       49           8
       50           4
       51           2
       52           1
       53           0
  163.7549850999785            125
FORTRAN STOP
```

References

[1] Talapov, A. L., Blöte, H. W.J., *J.Phys.* **A 29** 5727 (1996).
[2] Stauffer, D., Knecht, R., *Int.J.Mod.Phys.* **C 7** 893 (1996); Stauffer, D., *Physica* **A**, to be published (Widom Festschrift).
[3] Linke, A., Heermann, D. W., Altevogt, P., Siegert, M., *Physica* **A 222** 205 (1995).
[4] Koch, W., Dohm, V., Stauffer, D., *Phys. Rev. Lett.* **77** 1789 (1996).
[5] Sen, P., priv. comm.
[6] Stauffer, D., Aharony, A., *Introduction to Percolation Theory*, Taylor and Francis, (London, 1994); Sahimi, M., *Applications of Percolation Theory*, Taylor and Francis, (London, 1994); Bunde, A., Havlin, S., *Fractals and Disordered Systems*, (Springer, Heidelberg, 1996).
[7] Ziff, R. M., *Phys. Rev. Lett.* **69** 2670 (1992); see also Vicsek, T., Kertész, J., *Phys. Lett.* **81 A** 51 (1981).
[8] Newman, C. M., Schulman, L. S., *J. Stat. Phys.* **26** 613 (1981).
[9] Aizenman, M., *Nucl. Phys.*, in press; Sen, P., *Int. J. Mod. Phys.* **C 7** 603 (1996) and **8** (1996); Hu, C.-K., Lin, C.-Y., *Phys. Rev. Lett.* **77** 8 (1996); see also de Arcangelis, L., *J. Phys.* **A 20** 3057 (1987).

Error Estimates on Averages of Correlated Data

Henrik Flyvbjerg[1,2]

[1] Höchstleistungsrechnenzentrum (HLRZ)
    Forschungszentrum Jülich, D-52425 Jülich, Germany
[2] Department of Optics and Fluid Dynamics
    Risø National Laboratory, DK-4000 Roskilde, Denmark

Abstract. We describe how the true statistical error on an average of correlated data can be obtained with ease and efficiency by a renormalization group method. The method is illustrated with numerical and analytical examples having finite as well as infinite range correlations.

1 Introduction

Computer simulations of physical systems by Monte Carlo methods or molecular dynamics typically produce raw data in the form of finite time series of *correlated data*. In a large number of cases, where *stationary* states are investigated, the first step in the data analysis consists in computing time averages. Since such averages are over *finite* times, they are *fluctuating* quantities: another simulation of the same system will typically give a different value for the same quantity. So the next step in the data analysis consists in estimating the *variance* of finite time averages. A mixed practice has developed around this problem.

A popular estimator for the error on a time average of correlated data is based on the correlation function for these data. There is actually a whole family of such estimators, all being approximations to one of two original estimators. They are reviewed in Sect. 3 of this chapter with some attention paid to the approximations, subjective choices, and computational effort involved. That should make the reader appreciate the alternative, the "blocking" or "bunching," method, described in Sect. 4. In our opinion this method combines maximum rigor with minimum computation and reflection. It involves no approximations, nor subjective choices, automatically gives the correct answer when it is available from the time series being analyzed, and warns the user when this is not the case. We also give some—hopefully illustrative—examples, analytical ones (Sects. 5 and 6) as well as numerical ones (Sects. 7 and 8). In Sect. 9 we describe situations in which the simple blocking method cannot be used, and explain the jackknife method briefly, including a blocking method one can use with the jackknife. The reader who wants only a recipe for the blocking method, need only read Sects. 2, 4, and 9, and a few equations in Sect. 3.

The origin of the "blocking method" is unclear. It is part of the verbal tradition in part of the simulation community. It may have been invented by Ken Wilson [1]. This seems plausible, since it is essentially a real space renormalization group technique applied in the one-dimensional, discrete space of simulation time. The method was briefly described in several places (see e.g. [2], [3], [4], [5]). It was described and discussed in detail in [6], on which this chapter is based with some additions (and corrections of typographical errors). Since the publication of [6], other brief descriptions of the method have appeared, but none, to this author's knowledge, that gives the extensive and, hopefully, illuminating discussion given here.

2 Notation

Let $x_1, x_2, \ldots, x_n$ be the result of n consecutive measurements of some fluctuating quantity. Typically the x_i's may be the result of a Monte Carlo simulation in "thermal" equilibrium, i being related to the Monte Carlo time. Or the x_i's may be the result of a molecular dynamics simulation of a system in "equilibrium," i being related to the physical time of the system. Let $\langle \ldots \rangle$ denote the expectation value with respect to the exact, but unknown, probability distribution $p(x)$ according to which x_i is distributed. p does not depend on i, since we are considering a system in equilibrium. Let $\overline{\cdots}$ denote averages over the set $\{x_1, \ldots, x_n\}$. Then

$$\langle f \rangle \equiv \int dx\, p(x)\, f(x) \tag{1}$$

and

$$\bar{f} \equiv \frac{1}{n} \sum_{i=1}^{n} f(x_i) \ . \tag{2}$$

$\langle \cdots \rangle$ is good for theoretical considerations; $\overline{\cdots}$ is what we can compute in practice. We assume ergodicity; hence the ensemble average $\langle f \rangle$ is equal to the time average $\lim_{n \to \infty} \bar{f}$. In practice we compute finite time averages like (2), and use them as estimates for ensemble averages like (1). A finite average is a fluctuating quantity, so we need also an estimate for the *variance* of this quantity in order to have a complete result. To be specific, we estimate the expectation value $\mu \equiv \langle x \rangle$ by the average value

$$m \equiv \bar{x} = \frac{1}{n} \sum_{i=1}^{n} x_i \ , \tag{3}$$

and need an estimator for the variance of m,

$$\sigma^2(m) = \langle m^2 \rangle - \langle m \rangle^2 \ . \tag{4}$$

Inserting (3) into (4), we find that

$$\sigma^2(m) = \frac{1}{n^2} \sum_{i,j=1}^{n} \gamma_{i,j} = \frac{1}{n} \left[\gamma_0 + 2 \sum_{t=1}^{n-1} \left(1 - \frac{t}{n}\right) \gamma_t \right] \ , \tag{5}$$

where we have introduced the correlation function

$$\gamma_{i,j} \equiv \langle x_i x_j \rangle - \langle x_i \rangle \langle x_j \rangle \tag{6}$$

and used its invariance under time translations to define

$$\gamma_t \equiv \gamma_{i,j} \qquad t = |i - j| \ . \tag{7}$$

3 Estimators Based on Correlation Functions

$\sigma^2(m)$ is often estimated using (5) with an estimate for γ_t in γ_t's place. Doing so requires some care and consideration, since the most obvious estimator for γ_t,

$$c_t \equiv \frac{1}{n-t} \sum_{k=1}^{n-t} (x_k - \bar{x})(x_{k+t} - \bar{x}) \;, \tag{8}$$

is a *biased* estimator: its expectation value is *not* γ_t, but

$$\langle c_t \rangle = \gamma_t - \sigma^2(m) + \Delta_t \;, \tag{9}$$

where

$$\Delta_t = 2 \left(\frac{1}{n} \sum_{i=1}^{n} - \frac{1}{n-t} \sum_{i=1}^{n-t} \right) \frac{1}{n} \sum_{j=1}^{n} \gamma_{i,j} \;. \tag{10}$$

However, if the largest correlation time in γ_t is finite, call it τ, then (5) reads

$$\sigma^2(m) = \frac{1}{n} \left[\gamma_0 + 2 \sum_{t=1}^{T} \left(1 - \frac{t}{n}\right) \gamma_t \right] + \mathcal{O}\left[\frac{\tau}{n} \exp(-T/\tau)\right] \;, \tag{11}$$

where T is a cutoff parameter in the sum. For $\exp(-T/\tau) \ll 1$, the explicitly written terms in (8) clearly give a very good approximation to $\sigma^2(m) \sim \mathcal{O}(\tau/n)$. Furthermore, assuming $n \gg \tau$,

$$\Delta_t = \mathcal{O}\left(\frac{t\tau}{n^2}\right) \quad \text{for } t \ll \tau \;, \tag{12a}$$

growing to

$$\Delta_t = \mathcal{O}\left(\frac{\tau^2}{n^2}\right) \quad \text{for } t \gg \tau \;. \tag{12b}$$

So we may neglect Δ_t in (9), since it is at least a factor τ/n smaller than the term $\sigma^2(m) = \mathcal{O}(\tau/n)$. Doing that, and using (9) to eliminate γ_t from (5), we find that

$$\sigma^2(m) = \frac{1}{n} \left[\langle c_0 \rangle + 2 \sum_{t=1}^{T} \left(1 - \frac{t}{n}\right) \langle c_t \rangle \right] + \sigma^2(m) \left(\frac{1 + 2T}{n} - \frac{T(T+1)}{n^2} \right) \;. \tag{13}$$

Solving for $\sigma^2(m)$, we find

$$\sigma^2(m) \approx \left\langle \frac{c_0 + 2\sum_{t=1}^{T}(1 - \frac{t}{n})c_t}{n - 2T - 1 + \frac{T(T+1)}{n}} \right\rangle \;, \tag{14}$$

where the "$\approx$" is due to the truncation of the sum over t and the neglect of Δ_t. The expression inside the angular brackets in (14) is an estimator for $\sigma^2(m)$. This estimator is unbiased within the approximations done, as (14) shows, and it is *"the Mother of All Estimators"* based on the correlation function; see below.

Notice that with (14) we have not assumed anything about T's relation with n, except $T \leq n$. The truncated sum in (11) clearly approximates $\sigma^2(m)$ the better the larger T is, and is exact for $T = n-1$. The same is not true in (14). For $T = n - 1$, the denominator vanishes and the numerator is $\mathcal{O}(n\Delta_t) = \mathcal{O}(\tau^2/n)$. For T close to $n-1$, the right-hand side of (14) is the ratio of small numbers, one of which fluctuates, when $\langle c_t \rangle$ is estimated with c_t. This makes the expression inside the angular brackets in (14) a very bad estimator for $\sigma^2(m)$ for T close to $n - 1$. If instead one can choose T such that $\tau \leq T \leq n$, then the approximation in (11) is good, and so is that in (14). Thus T should at the same time be chosen much smaller than n and several times larger than the maximal correlation time in γ_t, assuming this quantity exists and can be determined at least approximately from c_t. See Sect. 7 for an example for which this is the case.

Alternatively, one may monitor the estimate for $\sigma^2(m)$ based on (14) as a function of T, and hope to demonstrate that it is independent of T for T larger than a certain value. This makes evaluation of c_t necessary for all values of t between 0 and $T_{\max}$, the maximal value for T considered. That requires $\mathcal{O}(nT_{\max})$ numerical operations, and easily becomes a time consuming affair. For this reason one would like to choose $T_{\max}$ small. On the other hand, $T_{\max}$ can in principle only be chosen appropriately *after* $\sigma^2(m)$ has been inspected. So the choice of $T_{\max}$ cannot be automatized. This lack of automatization and the computational effort involved are serious disadvantages of the method just described. In the next section we explain how the blocking method yields the desired result with only $\mathcal{O}(2n)$ numerical operations, and in a way that can be fully automatized.

The estimator given in Eq. (2.46) in [7], and in [8] and [9] is essentially (14), except the denominator is approximated by n. Daniell et al. [10] use the approximation

$$\sigma^2(m) \approx \left\langle \frac{c_0 + 2\sum_{t=1}^{T} c_t}{n - 2T - 1} \right\rangle \tag{15}$$

in their Eq. (23). One also sees the approximation

$$\sigma^2(m) \approx \left\langle \frac{c_0 + 2\sum_{t=1}^{T} c_t}{n} \right\rangle . \tag{16}$$

All these variants of (14) are equally good when T/n is sufficiently small. There is no reason not to use (14) itself, though, when any of the formulas are appropriate. It is as easy to compute as any of its approximations.

A variant of (8) in use is

$$\tilde{c}_t \equiv \frac{1}{n-t} \sum_{k=1}^{n-t} \left(x_k - \frac{1}{n-t} \sum_{k=1}^{n-t} x_k \right) \left(x_{k+t} - \frac{1}{n-t} \sum_{k=1}^{n-t} x_{k+t} \right) . \tag{17}$$

Like (8), (17) is a biased estimator for γ_t, since

$$\langle \tilde{c}_t \rangle = \gamma_t - \sigma^2(m) + \tilde{\Delta}_t , \tag{18}$$

where

$$\tilde{\Delta}_t = \left(\frac{1}{n^2} \sum_{i,j=1}^{n} - \frac{1}{(n-t)^2} \sum_{i=1}^{n-t} \sum_{j=t+1}^{n} \right) \gamma_{i,j} + \mathcal{O}\left(\frac{\tau t^2}{n^3}\right) . \tag{19}$$

Neglecting $\tilde{\Delta}_t$ relative to $\sigma^2(m)$ in (18) leads again to (13) and (14). Using (17) instead of (8) as estimator for $\langle c_t \rangle$ in (14) is a better approximation, when $|\sum_{t=1}^{T}(n-t)\tilde{\Delta}_t| < |\sum_{t=1}^{T}(n-t)\Delta_t|$, i.e., roughly when $T^2 < \tau n$.

4 The "Blocking" Method

We now describe a way to estimate $\sigma^2(m)$ which is computationally more economical than that of the previous section. It elegantly avoids the computation of c_t and the choice of T. In addition, it gives information about the quality of the estimate of $\sigma^2(m)$. The method involves repeated "blocking" of data, and computation of increasing lower bounds for $\sigma^2(m)$ in the following way.

We transform the data set $x_1, \ldots, x_n$ into half as large a data set $x'_1, \ldots, x'_{n'}$, where

$$x'_i = \tfrac{1}{2}(x_{2i-1} + x_{2i}) , \tag{20}$$
$$n' = \tfrac{1}{2}n . \tag{21}$$

We define m' as $\bar{x}'$, the average of the n' "new" data, and have

$$m' = m . \tag{22}$$

We also define $\gamma'_{i,j}$ and γ'_t as in (6) and (7), but from primed variables x'_i. One easily shows that

$$\gamma'_t = \begin{cases} \tfrac{1}{2}\gamma_0 + \tfrac{1}{2}\gamma_1 & \text{for } t = 0 \\ \tfrac{1}{4}\gamma_{2t-1} + \tfrac{1}{2}\gamma_{2t} + \tfrac{1}{4}\gamma_{2t+1} & \text{for } t > 0 \end{cases} \tag{23}$$

and that

$$\sigma^2(m') = \frac{1}{n'^2} \sum_{i,j=1}^{n'} \gamma'_{i,j} = \sigma^2(m) . \tag{24}$$

Equations (22) and (24) show that the two quantities we wish to know, m and $\sigma^2(m)$, are *invariant* under the "blocking" transformation given in (20,21). Thus, no desired information is lost in this transformation of the data set to half as large a set. Not only is nothing lost, but something is gained: the value of $\sigma^2(m)$ unravels gradually from γ_t by repeated application of the "blocking" transformation. From (5) we know that

$$\sigma^2(m) \geq \frac{\gamma_0}{n} , \tag{25}$$

and from (21) and (23) it is clear that γ_0/n increases every time the "blocking" transformation is applied, unless $\gamma_1 = 0$. In the latter case γ_0/n is invariant.

It is not difficult to show that $(\gamma_t/n)_{t=0,1,2,\ldots} \propto (\delta_{t,0})_{t=0,1,2,\ldots}$ is a *fixed point* of the linear transformation defined in (20,21), and any vector $(\gamma_t/n)_{t=0,1,2,\ldots}$ for which γ_t decreases faster than $1/t$ is in the basin of attraction of this fixed point. At this fixed point, $\sigma^2(m) = \gamma_0/n$ because $\gamma_t = 0$ for $t > 0$. $\sigma^2(m)$ is estimated by using (9) or (18) to eliminate γ_0 from (25), using $\Delta_0 = \tilde{\Delta}_0 = 0$ and solving for $\sigma^2(m)$:

$$\sigma^2(m) \geq \left\langle \frac{c_0}{n-1} \right\rangle, \qquad (26)$$

where c_0 is defined in (8), and the identity is satisfied at the fixed point. Knowing this, one proceeds as follows.

Starting with a data set $x_1, \ldots, x_n$, one computes $c_0/(n-1)$ and uses it as estimate for $\langle c_0 \rangle/(n-1)$. Then the "blocking" transformation (20,21) is applied to the data set, and $c'_0/(n'-1)$ is computed as estimate for $\langle c'_0 \rangle/(n'-1)$. This process is repeated until $n' = 2$. The sequence of values obtained for $c_0/(n-1)$ will increase until the fixed point is reached, whereupon it remains constant within fluctuations. The constant value is our estimate for $\sigma^2(m)$.

At the fixed point, the "blocked" variables $x'_1, \ldots, x'_{n'}$ are independent Gaussian variables—Gaussian by the central limit theorem, and independent by virtue of the fixed point value for γ'_t. Consequently, we can easily estimate the standard deviation on our estimate $c'_0/(n'-1)$ for $\sigma^2(m)$.

It is $\sqrt{2/(n'-1)}\, c'_0/(n'-1)$:

$$\sigma^2(m) \approx \frac{c'_0}{n'-1} \pm \sqrt{\frac{2}{n'-1}} \frac{c'_0}{n'-1}, \qquad (27)$$

$$\sigma(m) \approx \sqrt{\frac{c'_0}{n'-1}} \left(1 \pm \frac{1}{\sqrt{2(n'-1)}}\right). \qquad (28)$$

Knowing this error is a great help in determining whether the fixed point has been reached or not in actual calculations, as we shall see in Sect. 7.

If the fixed point is not reached before $n' = 2$, this is signaled by $c_0/(n-1)$ not becoming constant. The largest value obtained for $c_0/(n-1)$ is then a lower bound on $\sigma^2(m)$.

Notice that at no stage in these calculations were $(c_t)_{t=0,1,2,\ldots,T}$ evaluated, and the estimate for $\sigma^2(m)$ was obtained by $\mathcal{O}(2n)$ operations, while computation of $(c_t)_{t=0,1,2,\ldots,T}$ requires $\mathcal{O}(Tn)$ operations.

5 Analytical Example with Finite Correlation Time

Assume $x_1, \ldots, x_n$ are correlated with one finite correlation time, τ:

$$\gamma_t = \begin{cases} \sigma^2(x) & \text{for } t = 0 \\ A \exp(-t/\tau) & \text{for } t > 0 \end{cases} \qquad (29)$$

The "blocking" transformation maps the four parameters $(n, \tau, \sigma^2(x), A)$ into new values $(n', \tau', \sigma^2(x'), A')$, where

$$n' = \tfrac{1}{2}n$$
$$\tau' = \tfrac{1}{2}\tau$$
$$\sigma^2(x') = \tfrac{1}{2}(\sigma^2(x) + A\exp(\tau^{-1})) \tag{30}$$
$$A' = \tfrac{1}{2}(1 + \cosh\tau^{-1})A .$$

From (5) we get to leading order in $1/n$ (we do not have to make this approximation, but do it to keep formulas as simple as possible in this example) that

$$\sigma^2(m) = \frac{1}{n}\left(\sigma^2(x) + \frac{2A}{\exp(\tau^{-1}) - 1}\right) , \tag{31}$$

which is invariant under the "blocking" transformation (30). Using $\sigma^2(x)/n$ for $\langle c_0 \rangle/(n-1)$ in (26) and comparing with (31), we see that (26) underestimates $\sigma^2(x)$ by an amount δ, which can be inspected in the present example:

$$\delta = \frac{2A}{n(\exp(\tau^{-1}) - 1)} . \tag{32}$$

From (30) it follows that "blocking" gives

$$\delta' = \tfrac{1}{2}(1 + \exp(-\tau^{-1}))\delta , \tag{33}$$

from which we see that "blocking" makes $\langle c_0 \rangle/(n-1)$ grow to $\sigma^2(m)$ essentially with geometric progression. This is what one might have expected, knowing $\tau' = \tau/2$. However, this rate of convergence does *not* depend on τ being finite, as the example in the next section shows. It does not express a property of the time series being analyzed, but is due to the efficiency of the "blocking" algorithm.

6 Analytical Example with Infinite Correlation Time

Assume $x_1, \ldots, x_n$ are correlated with infinite correlation time, i.e., the correlation function decreases as a power of the difference in time:

$$\gamma_t = \begin{cases} \sigma^2(x) & \text{for } t = 0 \\ A/t^\alpha & \text{for } t > 0 . \end{cases} \tag{34}$$

Then (5) gives to leading order in $1/n$

$$\sigma^2(m) = \frac{1}{n}\left(\sigma^2(x) + 2A\zeta(\alpha)\right) \tag{35}$$

where $\zeta(\alpha)$ is Riemann's zeta function:

$$\zeta(\alpha) = \sum_{t=1}^{\infty} \frac{1}{t^\alpha}, \quad \text{Re}\,\alpha > 1 \ . \tag{36}$$

We see that if $\alpha < 1$ the data are too correlated to give a finite value for $\sigma^2(m)$. In general, the "blocking" transformation, (20,21), does not leave the form of the correlation function (34) invariant, since

$$\gamma'_t = \begin{cases} \frac{1}{2}(\sigma^2(x) + A) & \text{for } t = 0 \\ A/(2t)^\alpha \left(1 + \frac{\alpha(1+\alpha)}{16t^2} + \mathcal{O}(t^{-4})\right) & \text{for } t > 0 \ . \end{cases} \tag{37}$$

So this example is not entirely analytically tractable. However, in the most interesting situation, where α is not much larger than 1, $\sigma^2(m)$ in (5) receives a dominating contribution from γ_t with t large, as (35) and (36) show. For large t, γ_t is well approximated by the first term on the right-hand side of (37). This term *is* form invariant under the "blocking" transformation, giving

$$\begin{aligned} n' &= \tfrac{1}{2}n \\ \alpha' &= \alpha \\ \sigma^2(x') &= \tfrac{1}{2}\left(\sigma^2(x) + A\right) \quad \text{(not reliable)}, \\ A' &= 2^{-\alpha} A \ . \end{aligned} \tag{38}$$

As one would expect, (38) shows that "blocking" leaves the infinite correlation time infinite, and the power law unchanged, while the amplitude A is scaled in accordance with the power law.

Equation (38) is based upon an approximation which improves with distance t, so the relation between $\sigma^2(x')$ and $\sigma^2(x)$—quantities defined at zero distance—is not reliable. The relations for α and A are more reliable, and they are all we need. Comparing (35) with (25), remembering $\gamma_0 = \sigma^2(x)$, we see that $c_0/(n-1)$ underestimates $\sigma^2(m)$ by an amount δ, for which we have the explicit, approximate result

$$\delta = \frac{2A}{n}\zeta(\alpha) \ . \tag{39}$$

From (38) then follows that the "blocking" transformation gives

$$\delta' = 2^{1-\alpha}\delta \ , \tag{40}$$

i.e., δ vanishes geometrically, when we block transform it, even in this case of correlations with infinite range.

7 Numerical Example from Monte Carlo Simulation

We have simulated the two-dimensional Ising model on a 20×20 lattice at inverse temperature $\beta = 0.30$ using the heat bath algorithm and checkerboard update. After a hot start and 1000 thermalization sweeps we made 131072 ($= 2^{17}$) measurements of the instantaneous magnetization with consecutive measurements being separated by one sweep. With x_t denoting the instantaneous magnetization and $n \equiv 131072$, (3) gave $m = -0.0011$ for the magnetization.

We chose to measure the magnetization of this system to have a transparent example that is well known to most readers. The magnetization is easy to measure and even easier to discuss, since we know its exact value is zero. Consequently, our numerical estimate for the magnetization should differ from zero only by an amount that is justified by its error. Furthermore, with $\beta = 0.30$ the correlation time is short, and any method discussed above will work, so different methods can be compared easily.

We computed c_t as defined in (8). Figure 1 is a semilog plot of $c(t)$ versus t. For $t \leq 30$ we see a straight line, signaling a decrease of c_t with a single correlation time τ, which we read off the plot to be $\tau = 5.1$. We also read off $c_0 = 0.022$. It will be self-consistently correct to neglect $\sigma^2(m)$ in (9), and therefore to use c_t as estimator for γ_t. When (31) is used with $A = \sigma^2(x) = 0.022$, we find that $\sigma(m) = 0.0013$.

Figure 2 shows estimates for $\sigma(m)$ obtained from the same time series using the expression inside the angular bracket in (14) with increasing cutoff T as abscissa. From $15 \leq T \leq 300$ we see that $\sigma(m) = 0.0012$ independent of T. This constancy of T is a convincing signal that the value found for $\sigma(m)$ really *is* its true value. For $T > 300$ the values for $\sigma(m)$ become increasingly noisy. This is because c_t for $t \gg \tau$ is an essentially random number, and when T is increased, more such numbers are included in (14), giving rise to increasing noise.

Figure 3 shows estimates for $\sigma(m)$ obtained from the same time series and its block transformed series defined in (20) and (21). Equation (28) has been used. After approximately 6 block transformations $\sigma(m)$ reaches a value of 0.0012, where it remains (within error bars) during further block transformations. The distinct plateau that is seen in this plot of $\sigma(m)$ versus the number of block transformations applied, is a fully convincing signal that the fixed point for the block transformation has been reached, and $\sigma(m) = 0.0012$ is the true standard deviation on m. This value agrees with the previous estimate for it, and differs little from the estimate based on τ read off Fig. 1. It also makes our estimate $m = -0.0011$ for the magnetization differ less than one standard deviation from zero.

Now let us make a more detailed comparison of the results obtained with the two methods. Fig. 4 shows Figs. 2 and 3 plotted on top of each other. The abscissa of the points from Fig. 3 have been chosen such that $T = \frac{1}{2}(2^{\#}-1)$ where $\#$ is the number of block transformations applied. At this T value, the number of pair correlations $\overline{x_{t_1} x_{t_2}}$ taken into account by the estimate for $\sigma(m)$ based

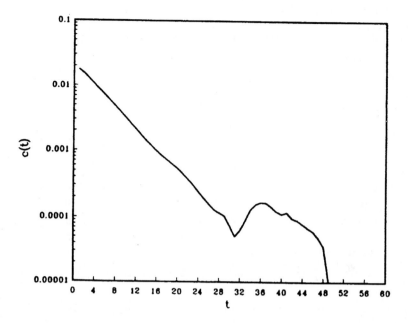

Fig. 1. c_t vs t. c_t was computed from $n = 2^{17}$ measurements of the magnetization $(x_t)_{t=1,\ldots,n}$ using the definition (8). For $0 \leq t \leq 16$, $c_t = 0.022 \exp(-t/5.1)$.

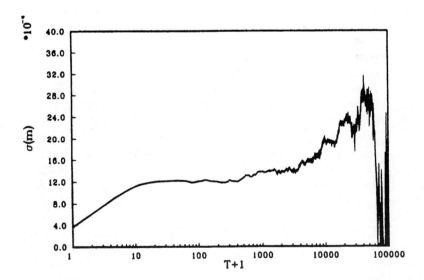

Fig. 2. Estimates for $\sigma(m)$ vs cutoff T, using (14). For $15 \leq T \leq 400$, $\sigma(m) \simeq 0.0012$ independent of T.

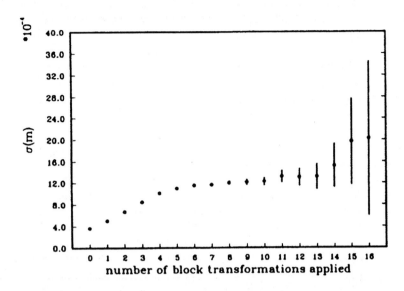

Fig. 3. Estimates for $\sigma(m)$ obtained with the blocking method. After approximately 6 block transformations, the estimates remain constant within error bars at 0.0012.

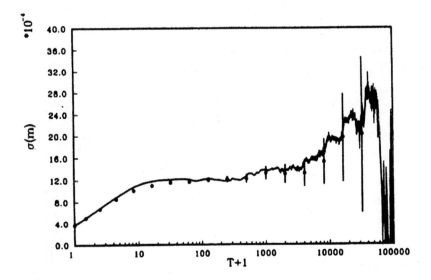

Fig. 4. Figures 2 and 3 plotted on top of each other. For reasons given in the text, the abscissa of the points from Fig. 3 have been chosen such that $T = \frac{1}{2}(2^{\#} - 1)$ where $\#$ is the number of block transformations applied.

on c_t is equal to the number of such pair correlations taken into account by the estimate for $\sigma(m)$ obtained with the blocking method. The *length* of the plateau giving this estimate is also the same for both methods. And the noise on the sequence of estimates obtained with one method stays within the error bars on the estimates obtained with the other method. So as far as the quality of results is concerned, both methods are equal, as they should be, since they both extract all relevant information from the time series. There is a great difference in the computational efficiency with which this is done, however. The data plotted in Fig. 2 were obtained with $\mathcal{O}(nT_{\max}) = \mathcal{O}(10^{10})$ arithmetical operations, whereas those in Fig. 3 required only $\mathcal{O}(n) = \mathcal{O}(10^5)$ operations. $T_{\max}$ is the efficiency ratio between the two methods, and equals 10^5 here only because we wanted to show the reader the noise in Fig. 2. Just the same, we could not have chosen $T_{\max}$ much less than 10^2, if we want to see the plateau in the estimate for $\sigma(m)$. Only if we *know* that $\tau \simeq 5$, can we legitimately choose $T_{\max}$ as low as $\simeq 20$ in (14). On the other hand, to obtain this knowledge we have to compute c_t at least for $0 \leq t \leq \max(\tau, 3)$, which requires $\mathcal{O}(n \max(\tau, 3))$ operations. In conclusion, there is no way to obtain $\sigma(m)$ which is more efficient than the blocking method, not even the crude method consisting in reading $c_0 \approx \sigma^2(x)$ and τ off Fig. 1 and using them in (28) with $A = \sigma^2(x)$.

Another difference between Figs. 2 and 3 is the lack of error bars in Fig. 2. If one can assume that $(x_t)_{t=1,\ldots,n}$ are *Gaussian* variables with correlation matrix γ_{t_1,t_2} (compare (6)) then these error bars be calculated from the estimate for γ_{t_1,t_2}. (Doing this is more than an exercise, we warn the reader.) The error bars in Fig. 3, on the other hand, are rigorous and independent of assumptions.

8 Numerical Example from Molecular Dynamics Simulation

The blocking method was tested on the results of a molecular dynamics run of 50,000 time steps of the Stockmeyer fluid, using $\mu^* = 1.5$, $T^* = 1.35$, $\rho^* = 0.8$, $N = 108$, where the symbols have their usual meaning [11]. This is an example with very long time correlations. The Ewald summation was used to evaluate the dipolar interactions in periodic boundary conditions, using a value of $\epsilon' = \infty$ for the dielectric constant of the surroundings [11], [12]. The same state point has been studied by Pollock and Alder (1980) [11], and our result for the Kirkwood factor g agrees with theirs. Their result for g was

$$g = \frac{1}{N\mu^2}\left(\sum_{i=1}^{N}\mu_i\right)^2 = 3.5 \pm 0.2 \ . \tag{41}$$

although they give no details of how their error estimate was arrived at.

The block transformation (2,21) was used to estimate the error in g. Where n was odd, one of the n measures of g was discarded before performing the next block transformation. The result of doing this is shown in Table 1. $\sigma(c_0/(n-1))$

was calculated using (28). A plateau value for $c_0/(n-1)$ is observed between 9 and 12 block transformations, and from this value we find $\sigma(g) \simeq \sqrt{0.024} \simeq 0.15$. In the last few rows in the table there are so few blocks that the error on $c_0/(n-1)$ is close to the value of $c_0/(n-1)$ itself. In [13], the blocking method was used to obtain error estimates for the Kirkwood factor in other simulations.

Table 1. Results of repeated application of the block transformation to a time series from a molecular dynamics simulation of the Stockmayer fluid.

Number of block transformations	n	$c_0/(n-1)$	$\sigma(c_0/(n-1))$
0	50 000	1.4×10^{-4}	
1	25 000	2.9×10^{-4}	
2	12 500	5.8×10^{-4}	
3	6 250	1.1×10^{-3}	
4	3 125	2.2×10^{-3}	
5	1 562	4.4×10^{-3}	
6	781	8.1×10^{-3}	
7	390	1.3×10^{-2}	
8	195	1.8×10^{-2}	0.2×10^{-2}
9	97	2.1×10^{-2}	0.3×10^{-2}
10	48	2.4×10^{-2}	0.5×10^{-2}
11	24	2.0×10^{-2}	0.6×10^{-2}
12	12	2.0×10^{-2}	0.9×10^{-2}
13	6	1.6×10^{-2}	1.0×10^{-2}
14	3	1.2×10^{-2}	1.2×10^{-2}

9 The Jackknife Method, and Its "Block" Transformation

There are situations in which the "blocking" method does not apply directly in the form presented here. Suppose, for example, that we wish to compute the spatial correlation length, ξ, for a spin model on a lattice—it could be the Ising model at a non-critical temperature—and we proceed to do that by Monte Carlo simulation, i.e., by generating a finite time series of correlated spin configurations,

$$(s_i(\mathbf{x}))_{\mathbf{x} \in \text{lattice}}, \quad i = 1, 2, \ldots, n \ . \tag{42}$$

The correlation length is defined from the asymptotic large-distance behavior of the spin-spin correlation function, γ,

$$\gamma(\mathbf{r}) = \langle s(\mathbf{x}) s(\mathbf{x}+\mathbf{r}) \rangle - \langle s(\mathbf{x}) \rangle^2 \propto \exp(-r/\xi) \ , \tag{43}$$

where $r = |\mathbf{r}|$ (but ξ's value typically depends on $\mathbf{r}$'s angle with the lattice axis).

We can easily obtain a series of estimates, $c_i(\mathbf{r})$, $i = 1, 2, \ldots, n$, for the spatial correlation *function*, $\gamma(\mathbf{r})$. We can estimate $\gamma(\mathbf{r})$ at Monte Carlo time i by the spin-spin correlations in the ith spin configuration,

$$c_i(\mathbf{r}) = \overline{s_i(\mathbf{x}) s_i(\mathbf{x}+\mathbf{r})} - \overline{s_i(\mathbf{x})}^2 \ , \tag{44}$$

where the bar denotes averaging over the lattice w.r.t. **x**. In order to obtain a corresponding series of values for the correlation *length*, however, we must take the logarithm of our estimates for the correlation function, and that at large distances, where the function has exponentially small positive values. Such estimates can take negative values, if they are not very precise, and that is typically the case when they have been obtained by spatial averaging over a single spin configuration, as here. When they do take negative values, we are left without even a rough estimate for the correlation length. Thus, a *good* estimate for the correlation function is imperative to obtain *any* estimate for the correlation length. For this reason, one typically *first* averages the estimates $c_i(\mathbf{r})$ over simulation time i, and only *then* takes its logarithm to obtain an estimate for the correlation length. But, as we saw, application of the blocking method to the correlation length seems to require the impossible: *first* taking the logarithm of $c_i(\mathbf{r})$ to obtain a series ξ_i, $i = 1, 2, \ldots, n$, for the correlation length, and *then* do the averaging and blocking.

This problem is a general one, arising whenever one wishes to calculate a function, $f[\cdot]$, of an expectation value, μ, whose estimator, m, can yield values beyond the range of definition of the function. So what to do? How to proceed instead?

First, note that we assume we have sufficiently good statistics for $\bar{x}$ to fall in the range of $f[\cdot]$. For example, when we estimate the correlation function, γ, by averaging over *all* spin configurations of the simulation, then the estimate, $\bar{c}$, is positive where we need its value, so we have no problem estimating the correlation length $\xi = f[\bar{c}]$ itself. The difficulty is in obtaining an error bar on this estimate for the correlation length – or, in the general case, for $f[\bar{x}]$.

Here is how to proceed [4], in the notation of the general case: Calculate the n estimates x_i as before, but then form n *other* estimates from them by averaging over all but the j'th one:

$$\tilde{x}_j = \frac{1}{n-1} \sum_{i \neq j} x_i \tag{45}$$

While the series $(f[x_i])_{i=1,\ldots,n}$ typically does not exist, the series $(f[\tilde{x}_i])_{i=1,\ldots,n}$ typically does, because $\tilde{x}_i$ differs little from $\bar{x}$, compare (3) and (45). Now use the estimate:

$$\sigma^2\left(f[\bar{x}]\right) \approx (n-1) \cdot \frac{1}{n} \sum_{i=1}^{n} (f[\tilde{x}_i] - f[\bar{x}])^2 \ . \tag{46}$$

In the simple case where $f[x] = x$, the right hand side of (46) is $c_0/(n-1)$, and (46) is just the estimate (27), except we have not done any blocking so far, so (46) is an underestimate when data are correlated.

A series of increasingly better estimates for $\sigma^2\left(f[\bar{x}]\right)$ is produced by *leaving out* not just x_j in the definition of $\tilde{x}_j$, but a whole block of m consequtive values,[1]

$$\tilde{x}_1 = \frac{1}{n-m} \sum_{i=m+1}^{n} x_i$$

$$\vdots$$

$$\tilde{x}_j = \frac{1}{n-m} \left(\sum_{i=1}^{(j-1)m} + \sum_{i=jm+1}^{n} \right) x_i \qquad (47)$$

$$\vdots$$

$$\tilde{x}_{n/m} = \frac{1}{n-m} \sum_{i=n-m+1}^{n} x_i \;,$$

with $m = 1, 2, 4, 8, \ldots, n/2$. Clearly, the number of values x_i shared between different $\tilde{x}_j$s decreases as m is increased, and reaches zero for $m = n/2$. So it should be clear, at least intuitively, that this procedure does the job. Extensive discussions are found, e.g., in [14], [15], [16], [17]. There one also finds discussions of how to get rid of possible bias: Even if $\bar{x}$ is an unbiased estimator, $f[\bar{x}]$ typically is not: If $\bar{x}$ is an unbiased estimator for just a real number, and f is a real function, then, with $\langle \bar{x} \rangle = \mu$ as above,

$$\langle f[\bar{x}] \rangle = f[\mu] + \tfrac{1}{2} f''[\mu] \sigma^2(\bar{x}) + \ldots \; . \qquad (48)$$

There is a cure for this as well, but we stop here.

10 Conclusion

I hope the reader to whom the blocking method was new, has learned enough from this presentation to apply it with confidence. Maybe even readers who use the method routinely may have learned a little from the examples. The main message was that no other method is comparable with the blocking method in computational and intellectual economy.

Acknowledgments

Like so many others, I learned the "blocking" method in a few sentences, from Enzo Marinari, I believe it was. The analytical examples given here, I worked out for a course on Monte Carlo simulation that I gave at Odense University, Denmark, in 1985. Later, a student from this course gave a talk at a conference, and received much attention for the way he had calculated the *error bars* on his results. Thus we learned that the "blocking method" was unknown to a section of

---

[1] This integer m should not be confused with the magnetization m, nor the estimator m for μ, used above. All these uses of the notation m are conventional.

the simulation community. Urged to do so, we wrote [6]. I thank Henrik Gordon Petersen for his permission to use his material here, and the audience at the *Eötvös Summer School in Physics* for being a second set of guinea pigs for my explanations. Their questions have improved the presentation of the material. I thank also the organizers of the school for the opportunity to present the material there.

References

[1] Wilson, K., *Monte Carlo calculations for the lattice gauge theory*. In 't Hooft et al., G., editor, *Recent Developments in Gauge Theories*, (Cargése, 1979. Plenum, New York, 1980).
[2] Whitmer, C., *Phys. Rev.*D **29** 306 (1984).
[3] Jacucci, G. and Rahman, A., *Nuovo Cimento* **D 4** 341 (1984).
[4] Gottlieb, S., Mackenzie, P. B., Thacker, H. B., Weingarten, D., *Nucl. Phys.* **B 263** 704 (1986).
[5] Allen, M. P. and Tildesley, D. J., *Computer simulation of liquids*. (Clarendon Press, Oxford. 1987).
[6] Flyvbjerg, H. and Petersen, H. G., *J. Chem. Phys.* **91** 461 (1989).
[7] Binder, K., Monte Carlo investigations. In Domb, C. and Green, M., editors, *Phase Transitions and Critical Phenomena*, volume 5. (Academic, New York, 1976).
[8] Binder, K., Theory and "technical" aspects of Monte Carlo simulations. In Binder, K., editor, *Monte Carlo Methods in Statistical Physics*, volume 7 of *Topics in Current Physics*. (Springer-Verlag, New York, 1979, 2nd ed. 1986).
[9] Binder, K. and Stauffer, D., A simple introduction to Monte Carlo simulation and some specialized topics. In Binder, K., editor, *Applications of the Monte Carlo Method in Statistical Physics*, volume 36 of *Topics in Current Physics*. (Springer, New York, 1984, 2nd ed. 1987).
[10] Daniell, G. J., Hey, A. J. G., and Mandula, J. E., *Phys. Rev.* **D 30** 2230 (1984).
[11] Pollock, E. L. and Alder, B. J., *Physica* **A 102** 1 (1980).
[12] de Leeuw, S. W., Perram, J. W., and Smith, E. R., *Proc. R. Soc. London Ser.* **A 373** 27 (1980).
[13] Petersen, H. G., de Leeuw, S. W., and Perram, J. W., *Mol. Phys.* **66** 637 (1989).
[14] Miller, R. G., *Biometrika* **61** 1 (1974).
[15] Efron, B., *SIAM Review* **21** 460 (1979).
[16] Efron, B., *The Jackknife, the Bootstrap and Other Resampling Plans*, volume 38 of *Regional Conference Series in Applied Mathematics*. (SIAM, 1982).
[17] Efron, B. and Tibshirani, R. J., *An introduction to the bootstrap*. (Chapman & Hall, London, 1993).

Stochastic Differential Equations

Dietrich E. Wolf

Gerhard-Mercator-Universität, FB 10, D-47048 Duisburg, Germany

Abstract. Elementary concepts of stochastic differential equations (SDE) and algorithms for their numerical solution are reviewed and illustrated by the physical problems of Brownian motion (ordinary SDE) and surface growth (partial SDE). Discretization schemes, systematic errors and instabilities are discussed. For surface growth also some recent results are presented.

1 Brownian Motion and Some General Remarks

Brownian motion is the best known and simplest example of a physical process which can be described by an ordinary stochastic differential equation (OSDE) such as

$$\frac{dh}{dt} = f(h(t)) + \eta(t). \qquad (1)$$

The time evolution of the degree of freedom $h(t)$, which could be one coordinate of the Browninan particle, contains a deterministic term, $f(h)$, and a stochastic contribution, η, called noise. Physical examples will be given below, but one concept should be made clear from the outset: Whereas the ordinary differential equation (ODE)

$$\frac{dh}{dt} = f(h(t)) \qquad (2)$$

specifies a single trajectory for given initial condition $h(t = 0)$, (1) describes a whole *ensemble of trajectories*, each of which is determined by a certain realization of the noise. Therefore, (1) only makes sense, if one specifies in addition to the initial conditions the statistical properties of the noise also, i.e. its probability distribution $P(\eta)$, or at least the moments needed for the ensemble averages one is interested in.

More generally, the functions $f(h)$ and $P(\eta)$ may also depend on the time t explicitly, which means that the deterministic evolution law or the properties of the noise change in time. This possibility will not be taken into account in the following, in order to keep the equations simple. The generalization is straight forward.

Excellent textbooks on this subject have been written, e.g. [1] or [2]. Numerical methods of solving stochastic differential equations can be found in the textbooks [3] and [4]. The present lecture notes are a 4 hour course, focusing on what I view as the essential concepts and tools, and leading to the most recent investigations of stochastic differential equations with a computer.

1.1 The Langevin equation

In 1827 the biologist R. Brown observed that a small particle dispersed in a liquid is in constant irregular motion, and he showed that this is a general physical property and not related to the particle being alive. The correct physical explanation dates back to F. M. Exner who proposed in 1900 that this Brownian motion is caused by random collisions of liquid molecules with the particle at a high rate due to their thermal motion. The real breakthrough in understanding this phenomenon was due to A. Einstein (1905) and M. von Smoluchowski (1906) who worked out the stochastic theory of Brownian motion. They described it by a deterministic differential equation (Focker-Planck-equation) for the probability distribution of the particle position. In 1908 P. Langevin introduced the concept of stochastic differential equations and showed that this intuitively more appealing description reproduces all the physical results derived by Einstein and Smoluchowski. Since then stochastic differential equations have remained the most popular tool for modelling stochastic processes in nature.

Langevin formulated the equation of motion of a Brownian particle as a generalization of Newton's equation. Any velocity component $v(t)$ of the particle relative to the surrounding liquid far away is damped by Stoke's drag force $-6\pi\tilde{\eta}av(t)$, where $\tilde{\eta}$ is the viscosity of the liquid and a the particle radius. Newton's equation hence reads

$$m\frac{dv}{dt} = -6\pi\tilde{\eta}av(t),$$

with m denoting the particle mass. In addition to this average force the Langevin equation contains a fluctuating momentum transfer per unit time, $\eta_v(t)$, caused by the collisions of the liquid molecules with the particle:

$$m\frac{dv}{dt} = -6\pi\tilde{\eta}av(t) + \eta_v(t). \tag{3}$$

What is the probability distribution of the noise? As it describes deviations from the average acceleration, its average value is zero: $\langle\eta_v(t)\rangle = 0$. As $\eta_v(t)\Delta t$ is the sum of many momentum transfers of individual, statistically independent molecules in a short time interval Δt, it may be assumed to have a Gaussian distribution (central limit theorem). Furthermore, the fluctuations of the momentum transfer at different times $t \neq t'$ can be regarded as uncorrelated: $\langle\eta_v(t)\eta_v(t')\rangle = \langle\eta_v(t)\rangle\langle\eta_v(t')\rangle = 0$. Hence,

$$\langle\eta_v(t)\eta_v(t')\rangle = \Gamma\delta(t-t'). \tag{4}$$

This equation defines *white* Gaussian noise.

Equation (4) shows that, in general, the noise is not an ordinary function of t. The Dirac-δ is a generalized function defined, as usual, by its action when multiplied with an arbitrary analytic function and then integrated over a finite interval. We shall see, however, that knowledge of the moments of the noise suffices to calculate the statistical properties of the solution of the stochastic differential equation.

1.2 A further simplification: the random walk

Although the Langevin equation (3) is not difficult to solve and certainly needs no computers, it is instructive to simplify it even further – instructive both for illustrating the physics of Brownian motion as well as for laying out the basis for numerical algorithms.

The solution of the deterministic part of the Langevin equation (3) is

$$v = v_0 \exp(-t/\tau) \quad \text{with} \quad \tau = m/6\pi\tilde{\eta}a. \tag{5}$$

This shows that the velocity keeps a memory of any initial "kick" by the liquid molecules for a characteristic time τ. Also, in the presence of noise a positive velocity remains most likely positive over a time approximately equal to τ.

The velocity autocorrelation function is therefore expected to be

$$\langle v(t)v(t')\rangle = \frac{k_B T}{m} e^{-|t-t'|/\tau}, \tag{6}$$

where the equipartition theorem has been used:

$$\left\langle \frac{m}{2} v^2(t) \right\rangle = \frac{1}{2} k_B T. \tag{7}$$

Indeed, the solution of (3) has the correlation (6).

If we are only interested in the physics on time scales much larger than τ we may as well regard (6) as a regularized Dirac-δ and replace it by the approximation

$$\langle v(t)v(t')\rangle \approx 2\frac{k_B T}{m}\tau\delta(t-t'). \tag{8}$$

One easily checks that the autocorrelation functions (6) and (8) integrated over t' coincide. As according to (5) $\tau \propto m$, the approximation (8) formally corresponds to the limit $m \to 0$ of (6), and accordingly, within the same approximation, one would neglect the left hand side of (3). The physical reason is that on time scales larger than τ the accelerations must average out to zero.

As a result, for $v = dr/dt$, one obtains the simplest possible stochastic differential equation, the equation describing a random walk:

$$\frac{dr(t)}{dt} = \eta_r(t), \quad \text{with} \quad \eta_r = \eta_v/6\pi\tilde{\eta}a. \tag{9}$$

According to (8), (5) the correlator of the noise is given by

$$\langle \eta_r(t)\eta_r(t')\rangle = 2D\delta(t-t'), \tag{10}$$

where the abbreviation

$$D = k_B T/6\pi\tilde{\eta}a \tag{11}$$

has been introduced.

The solution of (9) is

$$r(t) = r(0) + \int_0^t \eta_r(t')dt'. \tag{12}$$

Obviously this is not a function, but for each realization of the noise $\eta_r(t)$ a different (generalized) function: Equation (12) describes an *ensemble of trajectories*. The average position of the particle is

$$\langle r(t) \rangle = r(0) + \int_0^t \langle \eta_r(t') \rangle dt' = r(0), \tag{13}$$

where $\langle \eta_r \rangle = 0$ has been used. The mean square displacement is calculated using (10):

$$\langle (r(t) - r(0))^2 \rangle = \int_0^t \int_0^t \langle \eta_r(t_1)\eta_r(t_2) \rangle dt_1 dt_2 = 2Dt. \tag{14}$$

This shows that D has the physical meaning of the diffusion constant, and (11) is the famous Einstein relation expressing its dependence on temperature and viscosity of the surrounding liquid.

1.3 What have we learnt so far?

1. The example of Brownian motion illustrates that stochastic differential equations in physics are the result of an elimination of a huge number of microscopic degrees of freedom (the "bath", here the positions and velocities of liquid molecules) in order to focus on a small number of macroscopic degrees of freedom.
2. The noise is often assumed to be white Gaussian noise. This is always an idealization, valid only on time scales larger than the microscopic correlation times.
3. Whereas the solution of a deterministic differential equation for fixed initial condition is a trajectory, the solution of a stochastic differential equation is an ensemble of trajectories created by the ensemble of different realizations of the noise.

Consequently the solution of a stochastic differential equation on a computer means that one has to obtain a *representative sample of approximate trajectories* for different realizations of the noise. In order to assess the quality of the results one has to check two things:
a) How good are the approximate trajectories?
b) How representative is the sample?
The first question concerns mainly systematic errors of the integration scheme. As will be seen in the following sections, its answers are specifically different for stochastic and deterministic differential equations. By contrast the second question concerns statistical errors, which are conceptually the same as in any Monte-Carlo method. Improvement here usually requires an enlargement of the

sample: Since the probability distribution of the ensemble is often not known for nonequilibrium systems, the more sophisticated methods of choosing better sample generating algorithms (see e.g. the lectures of W. Krauth and F. Niedermayer in this course) are not applicable here.

2 Time Discretization

2.1 Euler algorithm for deterministic ODEs

In order to solve the ordinary differential equation (2) with initial condition $h(0) = h_0$ on a computer, one has to discretize time: Computers cannot calculate a continuum of function values. The aim is to calculate approximately $h(t_n)$ for the times

$$t_n = n\Delta t, \qquad n = 1, 2, ..., N, \qquad (15)$$

where $t_N = T$ is the maximum time, at which one wants to know the solution. The approximate values are denoted by $H(t_n)$, and it is important to assess the systematic error of this approximation.

Suppose we already have an approximate value $H(t_n) \approx h(t_n)$. The Euler scheme gets $H(t_n + \Delta t)$ by linear extrapolation. The Taylor–expansion

$$h(t + \Delta t) - h(t) = \int_t^{t+\Delta t} \frac{dh}{dt}(t') dt'$$

$$\approx f(h(t))\Delta t + \mathrm{O}(\Delta t^2)$$

is used to get, by inserting the approximate value at t_n,

$$H(t_{n+1}) = H(t_n) + f(H(t_n))\Delta t + \mathrm{O}(\Delta t^2). \qquad (16)$$

$H(t_n)$ has an error which accumulated over the previous steps $t_1, ... t_n$. By induction one sees that this error is of the order $n(\Delta t)^2$. Therefore inserting the approximate value of $h(t_n)$ in the function f gives rise to an extra contribution to the single step error which is of order $\mathrm{O}(n(\Delta t)^3)$, and can thus be subsumed into the error which was already made in the Taylor expansion.

The single step errors $\mathrm{O}(\Delta t^2)$ accumulate to a total systematic error

$$H(T) - h(T) = \mathrm{O}(\Delta t). \qquad (17)$$

This is a simple sum of the $N = T/\Delta t$ single step errors. We shall not discuss rounding errors here. They can pragmatically be regarded as part of the noise in the stochastic differential equation.

2.2 Stability of the Euler-scheme

The discussion at the end of the previous section tells us how the systematic errors vanish if one lets $\Delta t \to 0$. In practice, one cannot take this limit, of course, because the compution time for $h(T)$ would diverge. Therefore the question arises, what is the largest affordable Δt such that the approximate numerical solution still reflects the real time evolution? This question can be answered by considering an exponential relaxation process with decay time τ, $dh/dt = -h/\tau$, which has the solution $h(t) = h_0 \exp(-t/\tau)$. The Euler–scheme $H(t_{n+1}) = H(t_n)(1 - \Delta t/\tau)$ has the solution

$$H(t_n) = h_0 \left(1 - \Delta t/\tau\right)^n,$$

which has to be compared with the exact solution

$$h(t_n) = h_0 \left(e^{-\Delta t/\tau}\right)^n.$$

One sees that for $\Delta t \ll \tau$ the approximation is faithful to the time evolution. For $\tau < \Delta t < 2\tau$ the approximation oscillates but still reproduces the asymptotic value 0. Although the time evolution is no longer correct, the numerical scheme is called stable, as H remains bounded. However, for $\Delta t > 2\tau$ the Euler scheme becomes unstable: H oscillates with exponentially increasing amplitude.

This result can easily be generalized: If one has a spectrum of relaxation times, the shortest one determines the affordable discretization interval Δt. This is the essence of the von Neumann stability analysis (see e.g. [5], p.625).

2.3 The Wiener process

If one wants to apply the same type of reasoning, which lead to (16), to the case of an OSDE (1),

$$dh/dt = f(h(t)) + \eta(t)$$

with

$$\langle \eta \rangle = 0, \quad \langle \eta(t)\eta(t') \rangle = 2D\delta(t - t'),$$

one has to consider

$$\Delta w(t_n) = \int_{t_n}^{t_n + \Delta t} \eta(t')dt'. \tag{18}$$

Δw is a stochastic variable, the increment of the Wiener process. This we already know from (12): The position of a Brownian particle is a Wiener process. Hence, the $\Delta w(t_n)$ are statistically independent random variables with a Gaussian distribution [1] and

$$\langle \Delta w(t_n) \rangle = 0, \quad \langle \Delta w(t_n) \Delta w(t_{n'}) \rangle = 2D\Delta t \delta_{nn'}, \tag{19}$$

---

[1] This has not been proven here, but follows either by calculating the moments $\langle (\Delta w)^n \rangle$ or from the fact that the probability distribution of the position of a Brownian particle is a solution of the diffusion equation $\partial p(w)/\partial t = D\partial^2 p/\partial w^2$.

where $\delta_{nn'}$ is the Kronecker symbol:

$$\delta_{nn'} = \begin{cases} 1 \text{ if } n = n' \\ 0 \text{ otherwise} \end{cases}.$$

(19) implies that the increment of the Wiener process is of the order $\sqrt{\Delta t}$; hence, in contrast to the discretization of the deterministic equation, the stochastic one contains terms of order $\sqrt{\Delta t}$.

2.4 Euler algorithm for OSDEs

Applying the same type of reasoning as in (16) one can now write down the Euler approximation for a stochastic differential equation

$$\frac{dh}{dt} = f(h(t)) + \eta(t)$$

by collecting all terms up to order Δt. However, one has to be careful when estimating the systematic error. The lowest order term in the change of h over the time interval Δt is the Wiener increment $\Delta w \sim O(\sqrt{\Delta t})$. Expanding the integrand in

$$h(t_n + \Delta t) - h(t_n) = \int_{t_n}^{t_n + \Delta t} f(h(t'))dt' + \Delta w(t_n)$$

around $h(t_n)$ one therefore has $f(h(t_n)) + O(\sqrt{\Delta t})$. Thus one obtains

$$h(t_n + \Delta t) - h(t_n) = f(h(t_n))\Delta t + \Delta w(t_n) + O(\Delta t^{3/2}).$$

If we already know the approximation $H(t_n)$ of $h(t_n)$, the Euler scheme estimates $H(t_{n+1})$ by

$$H(t_{n+1}) = H(t_n) + f(H(t_n))\Delta t + \Delta w(t_n) + O(\Delta t^{3/2}). \tag{20}$$

The single step error is now larger than for the deterministic equation: $O(\Delta t^{3/2})$ instead of $O(\Delta t^2)$. It accumulates to a total systematic error of the trajectory:

$$H(T) - h(T) = O(\sqrt{\Delta t}). \tag{21}$$

What does the error of a stochastic trajectory mean for the error of ensemble averages? If we want to estimate the expectation value of any power of h, $\langle h^n(t) \rangle$ or $\langle h(t_1)...h(t_n) \rangle$, then in fact the systematic error is only $O(\Delta t)$. The reason is simply that the error order $O(\sqrt{\Delta t})$ is accumulated from terms of odd order in Δw, hence their average is zero.

On the other hand, if one measures a specific property of each trajectory, e.g. the time at which it first reaches a prescribed value, and averages this quantity afterwards to get e. g. the mean first passage time, then one needs special tricks if one wants to avoid errors of order $O(\sqrt{\Delta t})$ [6]. Knowledge of the dependence of the systematic error on Δt often allows one to extrapolate to $\Delta t = 0$ from measurements for several Δt.

2.5 How to simulate the Wiener process

In (20) the increments $\Delta w \equiv \sqrt{2D\Delta t}\, R$ (cf. (19)) of the Wiener process are drawn from a Gaussian distribution:

$$p(R) = \exp(-R^2/2)/\sqrt{2\pi} \qquad (22)$$

independently for each time step. This can be done by the Box-Muller algorithm described e.g. in [5], p.202. The following consideration [6] makes the simulation even more efficient (which is important as one needs a huge number of random numbers properly distributed):

As the Euler discretization scheme (20) is only accurate to order Δt one can afford to replace $\Delta w(t_n)$ by any other random variable $\Delta \tilde{w}(t_n)$ provided this does not introduce errors of order Δt. $\Delta \tilde{w}$ must have the same first and second moment as Δw. All higher moments are of higher order in Δt. Therefore one can set

$$\Delta \tilde{w}(t_n) = \sqrt{2D\Delta t}\, \tilde{R}_n \qquad (23)$$

where $\tilde{R}_n$ is uniformly distributed between $-\sqrt{3}$ and $\sqrt{3}$ so that $\langle \tilde{R}^2 \rangle = 1$ as in (22). With Rnd denoting a random number between 0 and 1, one can write

$$\tilde{R}_n = \sqrt{12}\,(\mathrm{Rnd}_n - 0.5). \qquad (24)$$

Sometimes one uses a discretization scheme of higher order than Euler. Then $\Delta w(t_n)$ has to be approximated with correct higher moments, consistently. Then an equidistribution does not suffice any more, and the Box-Muller algorithm becomes favorable.

2.6 On multiplicative noise

In contrast to what we have considered so far, often the noise is not additive. As an example let us consider the Verhulst equation from population dynamics:

$$\frac{dh}{dt} = \kappa h(t) - h^2(t). \qquad (25)$$

It describes the development of the size of a population in a habitat. The coefficient κ is the rate of exponential growth; the term $-h^2$ limits the population size and is due to the finite capacity of the habitat. It is plausible that the growth rate κ depends on stochastic environmental conditions: $\kappa = \bar{\kappa} + \eta(t)$. Then (25) is an example of a stochastic equation with multiplicative noise:

$$\frac{dh}{dt} = f(h(t)) + g(h(t))\eta(t) \quad \text{with} \quad g(h(t)) \neq \text{const.} \qquad (26)$$

Multiplicative noise gives rise to a mathematical problem known as the Itô-Stratonovich dilemma. This will be illustrated with a second example, which on first sight looks trivial: We transform the random walk equation (9),

$$\frac{dr}{dt} = \eta_r(t)$$

with a large initial value $r(0) = r_0 > 0$ into one for $r^2 \equiv X$:

$$\frac{dX}{dt} = 2r\frac{dr}{dt} = 2\sqrt{X}\,\eta_r(t). \tag{27}$$

This equation for the time evolution of X is correct as long as r remains positive. It is an equation with multiplicative noise. We obtained in Sect. 1 that

$$\langle X(t) \rangle = \langle r^2 \rangle = r_0^2 + 2Dt \tag{28}$$

(see (14), and we should reproduce this by using (27).

The Itô-discretization of (26) is

$$\Delta h = f(h(t))\Delta t + g(h(t))\Delta w, \tag{29}$$

while the Stratonovich scheme is implicit,

$$\Delta h = f(h(t) + \Delta h/2)\Delta t + g(h(t) + \Delta h/2)\Delta w. \tag{30}$$

For discussing both equations economically, we consider the interpolation

$$\Delta h = f(h(t) + \alpha \Delta h)\Delta t + g(h(t) + \alpha \Delta h)\Delta w. \tag{31}$$

Collecting all terms to order Δt in (31) one obtains

$$f(h + \alpha \Delta h)\Delta t = f(h)\Delta t + O(\Delta h \Delta t)$$
$$= f(h)\Delta t + O(\Delta t^{3/2}) \tag{32}$$
$$g(h + \alpha \Delta h)\Delta w = g(h)\Delta w + g'(h)\alpha \Delta h \Delta w + O(\Delta h^2 \Delta w)$$
$$= g(h)\Delta w + g'(h)\alpha g(h)\Delta w^2 + O(\Delta t^{3/2}) \tag{33}$$

Here we have used the fact that, up to terms of higher order in Δt, $\Delta w^2 \approx \langle \Delta w^2 \rangle = 2D\Delta t$, and $\Delta h = g(h)\Delta w + O(\Delta t)$. Inserting these results into (31) one finds that for multiplicative noise, i.e. for $g'(h) \neq 0$ the value of the interpolation parameter α makes a difference of order Δt:

$$\Delta h = f(h)\Delta t + g(h)\Delta w + \alpha(g^2(h))'D\Delta t + O(\Delta t^{3/2}). \tag{34}$$

Now we apply the general result (34) to the special case (27), where $f = 0$ and $g = 2\sqrt{X}$. The result is

$$\Delta X = \alpha\, 4D\Delta t + 2\sqrt{X}\,\Delta w. \tag{35}$$

In this equation, $\sqrt{X}$ is taken at the beginning of the interval Δt over which Δw is calculated. This means that the two factors are uncorrelated, and we can conclude that the expectation value of the last term in (35) is zero. Therefore we obtain $\langle \Delta X \rangle = \alpha 4D\Delta t$, or after summation over all time intervals up to t:

$$\langle X(t) \rangle = r_0^2 + \alpha\, 4Dt. \tag{36}$$

Comparision with (14) shows that the Stratonovich discretization scheme ($\alpha = 1/2$) gives the right answer.

The Wong-Zakai-Theorem (e.g. [3], p.84) tells us that this is generally true for all physical processes, where the white noise is only an idealization for negligible correlation time: The solutions of stochastic differential equations with multiplicative correlated noise, $\langle \eta(t)\eta(t') \rangle = A\exp(-|t-t'|/\tau)$ have the same statistical properties in the limit $\tau \to 0$, $A\tau = \text{const.}$, as the Stratonovich interpretation for white noise.

3 Surface Growth with Shot Noise

This example illustrates *partial* stochastic differential equations (PSDE). We shall see in Sect. 4 that the numerical treatment amounts to solving a set of coupled ordinary stochastic differential equations. The recent books [7] and [8] give overviews about surface growth.

3.1 Kinetic roughening

We consider the growth of a solid layer on a flat, d-dimensional substrate (surface dimensions $d > 2$ are only of theoretical interest). The layer thickness is denoted by $h(x,t)$, where x is the position on the substrate and t the duration of the growth process. For simplicity we assume that particles are deposited along trajectories perpendicular to the substrate at a rate F [particles per unit area and unit time]. The deposition rate fluctuates ("shot noise", see next section). Therefore an initially flat surface becomes rough during the growth process. This roughness is stronger than the thermal fluctuations of the surface. In order to distinguish it from the growth induced by the thermal roughening, one calls it *kinetic roughening*.

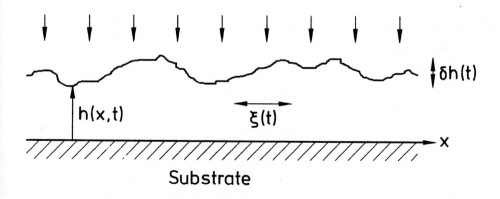

Fig. 1. When molecules are deposited to build a solid on a flat substrate, the surface becomes rough due to fluctuations in the deposition rate

After deposition time t the surface develops bumps up to wavelength $\xi(t)$ and amplitude $\delta h(t)$. It turns out that these scales depend on time algebraically:

$$\xi(t) \propto t^{1/z}, \tag{37}$$
$$\delta h(t) \propto t^{\zeta/z} \propto \xi(t)^\zeta \tag{38}$$

with the dynamic exponent z and the roughness exponent ζ (in the literature now most frequently denoted by χ). More precisely, we can measure the surface width

$$\delta h(L,t) \equiv \left\langle (h - \langle h \rangle)^2 \right\rangle^{1/2} \tag{39}$$

over a window of linear size L. Then

$$\delta h(L,t) = L^\zeta f(L/\xi(t)) \tag{40}$$

with

$$\begin{aligned} f(x) &= \text{const.} & \text{for} \quad L < \xi(t) \\ f(x) &\propto x^{-\zeta} & \text{for} \quad L > \xi(t). \end{aligned}$$

This scaling property of kinetic roughening was discovered by Family and Vicsek [9] and is called self-affinity to distinguish it from self-similarity, where $\zeta = 1$.

In order to classify the possible exponents ζ and z, one formulates stochastic differential equations for the time evolution of the film thickness. In the following we shall concentrate on the probably best known of these growth equations, the Kardar-Parisi-Zhang- or KPZ-equation ([10], see also [11]). Biological growth processes (e.g. tumor growth, or spread of epidemics) can also be modelled by this equation. Similar equations have recently also been proposed for the dynamics of granular media [12].

3.2 Shot noise

During a time interval t, the number of particles deposited independently at random positions and times in an area L^d is

$$FtL^d \pm (FtL^d)^{1/2}. \tag{41}$$

Its fluctuation is the shot noise. Therefore the thickness of the deposited layer is

$$\begin{aligned} h_L &= \left(FtL^d \pm (FtL^d)^{1/2} \right) / L^d \\ &= Ft \pm (Ft/L^d)^{1/2}. \end{aligned} \tag{42}$$

The index L denotes quantities which are spatial averages over an area L^d:

$$h_L(t) = L^{-d} \int_{L^d} h(x,t) \mathrm{d}^d x.$$

The $\sqrt{t}$-dependence in (42) shows that one can describe the shot noise as a Wiener process:

$$w_L(t) = \int_0^t \eta_L(t')dt' \tag{43}$$

with

$$\langle \eta_L(t)\eta_L(t')\rangle = FL^{-d}\delta(t-t'). \tag{44}$$

The shot noise is not only uncorrelated in time but also in space. Therefore its continuum description is $\eta(x,t)$ with $\langle \eta(x,t)\rangle = 0$ and

$$\langle \eta(x,t)\eta(x',t')\rangle = F\delta^d(x-x')\delta(t-t'). \tag{45}$$

One easily checks that (44) follows.

3.3 The Kardar-Parisi-Zhang (KPZ) equation

In many growth processes some of the deposited particles may be rejected from the surface, because they do not find a binding site, or the growing layer may have voids. Desorption and the formation of voids generically depend on the local inclination of the surface: E. g. on a crystal surface atoms stick to edges but may desorb from terraces. Or a tilted surface may develop overhangs which cause the formation of voids. In both cases the local height increase per unit time will depend on the surface inclination, ∇h.

For this type of growth processes, Kardar, Parisi and Zhang [10] proposed the following stochastic differential equation

$$\partial h(x,t)/\partial t = F + \nu\nabla^2 h + \lambda(\nabla h)^2 + \eta(x,t). \tag{46}$$

It contains a term $\nu\nabla^2 h$ which suppresses large curvature of the surface. The slope dependence of the growth velocity expressed by the λ-term is quadratic, because reflection symmetry $x \to -x$ is assumed. The constant F can be transformed away by going into a comoving frame, $h \to h - Ft$.

3.4 What is known about the scaling behaviour?

Before we discuss the algorithms used to solve e.g. (46) numerically, let us recall some analytical results which can be used for validating the numerical algorithm (see e.g. [11]).

The linear case, $\lambda = 0$, is exactly solvable by Fourier transformation. The roughness exponent ζ is $1/2$ for $d = 1$ and zero for $d > 2$. The dynamic exponent is $z = 2$, independent of d. At the upper critical dimension $d = 2$ the power law (38) is replaced by a logarithmic divergence $\delta h^2 \propto \log \xi$ with $\xi \propto t^{1/2}$.

For $\lambda \neq 0$, dynamic renormalization group calculations [13] show that the scaling behaviour is determined by the coupling constant

$$g \equiv F\lambda^2/\nu^3. \tag{47}$$

For $d = 1$ there are two fixed points: $g = 0$ corresponding to the linear theory is unstable, i.e. no matter how weak the nonlinearity is, on sufficiently large scales it will dominate the roughness. At the stable fixed point at $g \neq 0$ the exponents are given by $\zeta = 1/2$, $z = 3/2$.

For $d > 2$ one finds again two fixed points (in one-loop approximation). The one at $|g| = g_c > 0$ is now unstable and separates a weak coupling from a strong coupling regime. If $|g| < g_c$, the effective coupling on large scales goes to zero, so that the scaling is determined by the linear theory. However, if $|g| > g_c$, the coupling becomes stronger the larger the length scales one is interested in. Here, no exponents are known analytically. One can only show that $\zeta + z = 2$ in this regime.

For $d = 2$ the transition between strong and weak coupling happens already at $g_c = 0$. Any system with finite g is in the strong coupling phase, however the crossovers may last for very long, if g is small [14].

4 Space Discretization

4.1 Deterministic part

How to approximate functions and their derivatives using only the values $h(x_i) \equiv h_i$ on a space grid $x_i = i\Delta x$ is nowadays common knowledge taught in basic courses on numerics. The usual finite difference scheme gives the following expressions for the terms of the KPZ-equation:

$$\frac{\partial h}{\partial x}(x_i) = \frac{1}{2\Delta x}(h_{i+1} - h_{i-1}) + O(\Delta x^2), \tag{48}$$

which can be derived by subtracting the Taylor expansions for $h_{i\pm 1}$ from each other,

$$h_{i\pm 1} = h_i \pm \frac{\partial h}{\partial x}\Delta x + \frac{1}{2}\frac{\partial^2 h}{\partial x^2}(\Delta x)^2 \pm O(\Delta x^3). \tag{49}$$

Hence, the square gradient term of the KPZ-equation is approximated by

$$\left(\frac{\partial h}{\partial x}(x_i)\right)^2 = \frac{1}{4(\Delta x)^2}(h_{i+1} - h_{i-1})^2 + O\left(\Delta x^2\right). \tag{50}$$

Similarly, the second derivative is approximated by

$$\frac{\partial^2 h}{\partial x^2}(x_i) = \frac{1}{(\Delta x)^2}(h_{i+1} - 2h_i + h_{i-1}) + O(\Delta x^2) \tag{51}$$

(add the expansions (49)).

The generalization to higher dimensions is straightforward. After space discretization the PSDE is reduced to a set of coupled OSDE's:

$$\frac{dh_i}{dt} = \frac{1}{(\Delta x)^2}\left(\nu(h_{i+1} - 2h_i + h_{i-1}) + \frac{\lambda}{4}(h_{i+1} - h_{i-1})^2\right) + \text{noise}. \tag{52}$$

4.2 An unexpected intrinsic instability

The space discretization of the preceding section, which seems so straightforward, is intrinsically unstable, as discovered recently by Newman and Bray [15]. The instability has nothing to do with noise. It is caused by the nonlinearity of (46) and is related to the finite time singularities ("shocks") of the deterministic KPZ-equation for $\nu = 0$.

Consider the deterministic time evolution of a parabolic part of the surface, $h(x,0) = Kx^2/2$. It remains parabolic, but the curvature K changes. Inserting the Ansatz $h(x,t) = K(t)x^2/2$ into

$$\frac{\partial h}{\partial t} = \lambda \left(\frac{\partial h}{\partial x}\right)^2$$

one obtains

$$\frac{dK}{dt}\frac{x^2}{2} = \lambda K^2 x^2.$$

The solution

$$K(t) = \frac{K(0)}{1 - 2\lambda K(0)t} \tag{53}$$

diverges at

$$t = (2\lambda K(0))^{-1}. \tag{54}$$

Here, K and λ are assumed to have the same sign: For positive λ, the singularities occur at minima, for negative at maxima.

For $\nu \neq 0$ these singularities are suppressed. A stable situation forms in which the local curvature at the minimum is proportional to the slope squared, s^2, of the adjacent regions (see Fig. 2):

$$K = (\nabla^2 h)|_a \approx \frac{\lambda}{\nu}(\nabla h)^2|_b = \frac{\lambda}{\nu}s^2 \tag{55}$$

This guarantees that both parts grow with the same average velocity. It should be kept in mind that the deterministic part of the KPZ-equation does not change the slope of flat surface parts, so that s remains approximately constant, whereas the curvature at the minimum as well as the size of the curved region, x_0, adapts to it. Matching the slopes (a) of the curved region around the minimum, Kx_0, and (b) of the linear ones, s, at a distance x_0 from the minimum, one can estimate

$$x_0 \approx \frac{\nu}{\lambda s}. \tag{56}$$

Obviously, with a discrete space grid this stable situation cannot develop if the local slope exceeds $\nu/2\lambda\Delta x$, or using the discretization, if

$$|h_{i+2} - h_i| > \nu/\lambda, \tag{57}$$

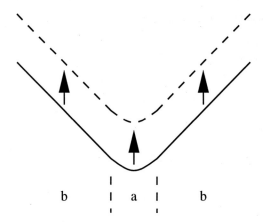

Fig. 2. For given slopes in region (b) the curvature at the minimum and the size of the curved region (a) adjust such that the surface grows everywhere with the same average speed

because then $x_0 < 2\Delta x$, according to (56). At the minimum the finite difference scheme of the curvature is then always too small, so that the surface lags behind more and more:

$$\nu \frac{1}{(\Delta x)^2}(h_1 - 2h_0 + h_{-1}) \approx \frac{2}{(\Delta x)^2}(h_1 - h_0)$$
$$< \frac{\lambda}{(\Delta x)^2}(h_2 - h_0)^2. \tag{58}$$

If somehow local slopes larger than $\nu/2\lambda\Delta x$ are produced, then a local minimum in the neighbourhood will stay back, hence the local slope increases even further, so that the surface becomes unstable. This example shows that the stability of the discretization of the deterministic part has to be carefully assessed before proceeding to the stochastic part and the numerical solution. In order to prevent the accidental occurence of such big slopes due to the noise, Moser, Kertész and Wolf [16] had to choose extremly small values of Δt. If the instability happened to occur nonetheless, that run had to be discarded.

Newman and Bray [15] devise a stable space discretization instead of (52):

$$\frac{dh_i}{dt} = \frac{1}{(\Delta x)^2}\left(\mathcal{F}(h_{i+1} - h_i) + \mathcal{F}(h_{i-1} - h_i)\right) + \text{noise} \tag{59}$$

with

$$\mathcal{F}(x) = \frac{\nu^2}{\lambda}\left(\exp\left(\frac{\lambda x}{\nu}\right) - 1\right). \tag{60}$$

Expanding this for small x to second order, the similarity to (52) becomes clear:

$$\mathcal{F}(x) \approx \nu x + \lambda x^2/2. \tag{61}$$

However, this scheme cannot be generalized into a recipe which is applicable to other nonlinear PSDE's.

4.3 Space discretization of the noise

One can view the introduction of a space grid as a coarse-graining procedure: One looks at the surface with a resolution Δx. No smaller features can be resolved. In this sense one also sees only the average effect of the noise on an interval Δx:

$$\eta_i(t) = \frac{1}{\Delta x} \int_{x_i - \Delta x/2}^{x_i + \Delta x/2} \eta(x,t) \mathrm{d}t. \tag{62}$$

One easily checks that the discretized noise is characterized by

$$\langle \eta_i(t) \rangle = 0,$$

$$\langle \eta_i(t) \eta_j(t') \rangle = \frac{F}{(\Delta x)^d} \delta_{ij} \, \delta(t-t') \tag{63}$$

(cf.(44)). This completes the spatial discretization of the KPZ-equation, which merely served as an example; the results can easily be adapted to other PSDE's.

4.4 Relaxation times of the discretized equation

With the results of Sect. 2 the time discretization of (52) with the noise (63) is now straightforward (Euler scheme) [16]:

$$\begin{aligned}H_i(t_{n+1}) &= H_i(t_n) + \frac{\Delta t}{(\Delta x)^2} \Big(\nu \left(H_{i+1}(t_n) - 2H_i(t_n) + H_{i-1}(t_n) \right) \\ &+ \frac{\lambda}{4} \left(H_{i+1}(t_n) - H_{i-1}(t_n) \right)^2 \Big) + \left(\frac{F \Delta t}{(\Delta x)^d} \right)^{1/2} \sqrt{12} \; (\mathrm{Rnd}_i - 0.5),\end{aligned} \tag{64}$$

where the surface dimension is $d = 1$. For higher dimensions, the deterministic part contains additional terms of the same form for the other spatial directions.

We have seen in Sect. 2.2 that the time discretization interval has to be smaller than the characteristic relaxation times of the system in order to assure numerical stability of the Euler scheme. For the linear equation, $\lambda = 0$, surface fluctuations with wave number $k = 2\pi/l$ emerge after time $\sim l^z$ with dynamic exponent 2. As none of these fluctuations grows forever, the production rate must be balanced by the relaxation rate:

$$\tau(l) \propto l^z. \tag{65}$$

Eqn.(65) holds also for $\lambda \neq 0$, but z is then smaller than 2. As the smallest wavelength compatible with the spatial discretization is Δx, the time increment Δt must be chosen such that

$$\Delta t < \mathrm{const.}(\Delta x)^z. \tag{66}$$

This requirement determines the affordable time step in all cases, where the instability described in Sect. 4.2 does not occur, in particular also in the discretization scheme (59) of [15].

4.5 Dimensionless formulation of the discretized equation: one more surprise

Eqn.(64) contains 5 dimensional parameters: ν, λ, F, Δx and Δt. Any three of them can be eliminated by choosing proper units for x, t and h. This means we can completely eliminate ν, λ and F from the equation and just retain the dimensionless discretization parameters $\Delta \tilde{x}$ and $\Delta \tilde{t}$. What determines then, whether or not we are in the strong or the weak coupling regime introduced in Sect. 3.4?

The natural units [16] are

$$h_0 = \frac{\nu}{\lambda}, \quad t_0 = \frac{\nu^2}{F\lambda^2}(\Delta x)^d, \quad x_0 = \left(\frac{\nu^3}{F\lambda^2}(\Delta x)^d\right)^{1/2}. \tag{67}$$

With $\tilde{H} = H/h_0$, $\tilde{t} = t/t_0$ and $\tilde{x} = x/x_0$ the discretized KPZ-equation now is dimensionless:

$$\tilde{H}_i(\tilde{t}_{n+1}) = \tilde{H}_i(\tilde{t}_n) + \frac{\Delta \tilde{t}}{(\Delta \tilde{x})^2}\left(\tilde{H}_{i+1}(\tilde{t}_n) - 2\tilde{H}_i(\tilde{t}_n) + \tilde{H}_{i-1}(\tilde{t}_n)\right)$$
$$+ \frac{1}{4}\left(\tilde{H}_{i+1}(\tilde{t}_n) - \tilde{H}_{i-1}(\tilde{t}_n)\right)^2 + \sqrt{12\Delta\tilde{t}}\ (\text{Rnd}_i - 0.5). \tag{68}$$

The coupling constant $g = \lambda^2 F/\nu^3$ (47) is not dimensionless. According to (67) it can be written in the form $g = (\Delta x)^d/x_0^2$, which most clearly shows that it has dimension $[length]^{d-2}$. Hence, the dimensionless coupling constant is

$$g/x_0^{d-2} = (\Delta \tilde{x})^d. \tag{69}$$

The surprising conclusion is that the spatial discretization determines whether or not one is in the strong coupling regime! The temporal discretization $\Delta \tilde{t}$ determines whether or not the discretization is numerically stable.

4.6 Concluding remarks and further reading

With the methods described in these lectures precise numerical values for the strong-coupling exponents of the KPZ-equation have been obtained for $d = 1, 2$ and 3 (see e.g. [16], [17]). Evidence for the transition between weak and strong coupling has been given for $d = 3$. Aspects of the implementation of the numerical integration schemes on vector and on parallel computers are discussed e.g. in [17].

If the square gradient term in (46) for $d = 2$ is more generally given by

$$\lambda_\parallel \left(\frac{\partial h}{\partial x_\parallel}\right)^2 + \lambda_\perp \left(\frac{\partial h}{\partial x_\perp}\right)^2, \tag{70}$$

as appropriate for vicinal surfaces [18], then the roughening is only logarithmic, if $\lambda_\parallel$ and $\lambda_\perp$ have opposite sign [19]. This prediction has also been confirmed by numerical solution of the stochastic differential equation (see e.g. [20],[21]). How to implement coloured noise is discussed e.g. in [22] (spatially correlated noise) and [23] (temporally correlated noise).

References

[1] Gardiner, C. W., *Handbook of Stochastic Methods*, (Springer, Berlin, Heidelberg, 1990).
[2] van Kampen, N. G., *Stochastic Processes in Physics and Chemistry*, (North Holland, Amsterdam, 1992).
[3] Honerkamp, J., *Stochastische Dynamische Systeme*, (VCH, Weinheim, 1990).
[4] Kloeden, P. E., Platen, E., Schurz, H., *Numerical Solution of SDE Through Computer Experiments*, (Springer, Berlin, Heidelberg, 1994).
[5] Press, W. H., Flannery, B. P., Teukolsky, S. A., Vetterling, W. T., *Numerical Recipes*, (Cambridge University Press, Cambridge, 1986).
[6] Seeßelberg, M., *Gewöhnliche und partielle Stochastische Differentialgleichungen*, (Diplomarbeit, Albert-Ludwigs-Univ. Freiburg, Fakultät für Physik, 1991).
[7] Barabási, A. L., Stanley, H. E. *Fractal Concepts in Surface Growth*, (Cambridge University Press, Cambridge, 1995).
[8] Villain, J., Pimpinelli, A., *Physique de la Croissance Cristalline*, (Éditions Eyrolles et Commissariat á l'Énergie Atomique, Paris, 1995).
[9] Family, F., Vicsek, T., *J. Phys.* **A 18** L75 (1985).
[10] Kardar, M., Parisi, G., Zhang, Y., *Phys. Rev. Lett.* **56** 889 (1986).
[11] Family, F., Vicsek, T., (eds.) *Dynamics of Fractal Surfaces*, (World Scientific, Singapore, 1991).
[12] Mehta, A., Luck, J. M., Needs, R. J., *Phys. Rev.* **E 53** 92 (1996).
[13] Medina, E., Hwa, T., Kardar, M., Zhang, Y.-C., *Phys. Rev.* **A 39** 3053 (1989).
[14] Tang, L.-H., Forrest, B. M., Wolf, D. E., *Phys. Rev.* **A 45** 7162 (1992).
[15] Newman, T. J., Bray, A. J., *J. Phys.* **A 29**, 7917 (1996).
[16] Moser, K., Kertész, J., Wolf, D. E., *Physica* **A 178** 215 (1991).
[17] Moser, K., Wolf, D. E., *J. Phys.* **A 27** 4049 (1991).
[18] Villain, J., *J. Phys.* **I 1** 19 (1991).
[19] Wolf, D. E., *Phys. Rev. Lett.* **67** 1783 (1991).
[20] Halpin-Healy, T., Assdah, A., *Phys. Rev.* **A 46** 3527 (1992).
[21] Moser, K., Wolf, D. E., Kinetic Roughening of Vicinal Surfaces; in: *Surface Disordering: Growth, Roughening, and Phase Transitions*, eds. R. Jullien, J. Kertész, P. Meakin, D. E. Wolf (Nova Science, Commack, 1992, p.21).
[22] Amar, J. G., Lam, P. M., Family, F., *Phys. Rev.* **A 43** 4548 (1991).
[23] Lam, C. H., Sander, L. H., Wolf, D. E., *Fractals* **1** 1 (1993).

Frustrated Systems: Ground State Properties via Combinatorial Optimization

Heiko Rieger

HLRZ c/o Forschungszentrum Jülich, 52425 Jülich, Germany

Abstract. An introduction to the application of combinatorial optimization methods to ground state calculations of frustrated, disordered systems is given. We discuss the interface problem in the random bond Ising ferromagnet, the random field Ising model, the diluted antiferromagnet in an external field, the spin glass problem, the solid-on-solid model with a disordered substrate and other convex cost flow problems occurring in superconducting flux line lattices and traffic flow networks. On the algorithmic side we present a concise introduction to a number of elementary algorithms in combinatorial optimization, in particular network flows: the shortest path algorithm, the maximum-flow algorithms and minimum-cost-flow algorithms. We also take a glance at the minimum weighted matching and branch-and-cut algorithms.

1 What Are Frustrated Systems?

Frustrated systems are simply systems in which the individual entities that build up the model (like spins, bosons, fermions, monomers, etc.) feel some sort of "frustration" in the literal sense. This means that on their search for a minimal energy configuration at lower and lower temperatures they are not able to satisfy all interactions with one another or with impurities simultaneously.

As an example we consider a model for a directed polymer in a disordered environment

$$H = \underbrace{\sum_i (x_i - x_{i+1})^2}_{(\mathbf{A})} - \underbrace{\sum_i V(x_i)}_{(\mathbf{B})}, \tag{1}$$

where x_i is the displacement of the i-th monomer and $V(x)$ is a random potential. The first term (A), the elastic energy, tries to make the polymer *straight* for $T \to 0$, the second term (B) tries to bring the monomers into favoured positions, for which the polymer has to *bend*. The monomers cannot satisfy both of these demands simultaneously.

Another, more famous example are magnetic spins (for simplicity Ising spins) with ferromagnetic *and* antiferromagnetic interactions. Consider the Hamiltonian for 4 spins (e.g. an elementary plaquette of a square lattice)

$$H = -\sigma_1\sigma_2 - \sigma_2\sigma_3 - \sigma_3\sigma_4 + \sigma_4\sigma_1 \tag{2}$$

and try to find a configuration of the Ising spins $\sigma_i = \pm 1$ that minimizes this simple energy function. Naively one starts with some value for the first spin, let's say $\sigma_1 = +1$, the first term would then imply $\sigma_2 = +1$, the second $\sigma_3 = +1$ and the third $\sigma_4 = +1$. But what about the last term – here $\sigma_1 = \sigma_4 = +1$ is *not* the most favorable configuration. Thus it is impossible to satisfy all local interactions at once, this is why Toulouse [1] introduced the concept of frustration for these plaquettes occurring naturally in spin glass models. After some thought one finds that many (i.e. 8) different spin configurations for (2) have a minimal energy, but all of them break one bond. This is the notorious frustration induced ground state degeneracy.

This kind of frustration can occur either via quenched disorder (i.e. a random, time-independent distribution of ferromagnetic and antiferromagnetic spin interactions) or without any disorder, for example in the fully frustrated antiferromagnetic Ising model on a triangular lattice. Of course the same problem occurs with XY-spins, like the XY-antiferromagnet on a triangular or Kagomé lattice. In this lecture we treat disorder induced frustration exclusively. The determination of ground states of regularly frustrated systems usually does not need such algorithmic tools as discussed in this lecture (see [2] and references therein for a number of examples).

2   What Is Special for Simulations of Disordered Systems?

As we learned from our simple 4-spin Hamiltonian above, frustration is often responsible for the existence of many degenerate (or nearly degenerate) states and metastable states. Suppose one intends to perform a conventional Monte Carlo simulation with the single spin flip heat bath dynamics of such a system. To explore the whole energy landscape in order to find the most favorable configurations one has to overcome large energy barriers between the various minima. As a consequence, the relaxation times typically become astronomically large — not only *at* a possible phase transition (in which case it would be "critical slowing down"), but also below *and* above. Thus, as is well known in the community of computational physicists (also among experimentalists, by the way), investigating disordered or amorphous materials, *equilibration* is nearly impossible for large systems. Our first commandment in this context is therefore:

> **To be modest in system size is mandatory!**

Of course everything would be fine if there were an efficient[1] cluster algorithm at hand, as discussed in this school. However, there are none — with a few exceptional cases. To invent an efficient cluster algorithm for some model, one first has to have a deep understanding of its physics and a knowledge or an intuition about the low lying energy configurations, the excitations etc. Thus, *cum grano salis*, if you have understood the system to a rather complete extent,

---

[1] We emphasize this word, because it is easy to formulate an algorithm that constructs *some* clusters. The question is, whether the flip acceptance rate is reasonable.

you might be ready to formulate a cluster algorithm — with which you can add some precise numbers (critical exponents etc.) to your basic understanding. Unfortunately, after several decades of research we still have not reached this desirable state for most of the interesting disordered systems.

The next observation is that different samples, i.e. different disorder realizations, can have completely different dynamical and static properties. This goes under the name "large sample to sample fluctuations", which originates in the lack of self averaging in some physical observables. Not all of them show this notorious behavior: the ground state energy, for instance, is well behaved, simply because the various local minima are nearly degenerate, but e.g. spatial correlation functions or susceptibilities are not self averaging quantities. Consider, for instance, the diluted ferromagnet with site concentration c and imagine the following two extreme situations: On a square lattice with N spins one could distribute $N/2$ spins in such a way that 1) they form a single, compact cluster, or 2) they occupy a sublattice such that none of them has an occupied nearest neighbor site. Obviously the magnetization or susceptibility has completely different characteristics in the two cases: 1) is a bulk ferromagnet of volume $N/2$ and will have a tendency to order ferromagnetically at some temperature T_c (in the limit $N \to \infty$), 2) is a collection of isolated spins that will never order.

Thus one easily recognizes that the probability distribution $P_L(\mathcal{O})$ of some observable $\mathcal{O}$ is usually extremely broad, in particular non-Gaussian. Rare events (i.e. disorder configurations with small probability) can have a strong impact on averaged quantities like susceptibilities or autocorrelations. This leads to our second commandment for *all* investigations of disordered systems:

> **Sample a huge (!) number of disorder configurations**

The study of the probability distribution $P_L(\mathcal{O})$ can be more useful than average values only.

3 What Can We Learn from Ground State Calculations?

The ground state is the configuration in which the "equilibrated" system settles at exactly zero temperature (if there is more than one, replace state by states). However, $T = 0$ is not accessible in the real world, so why should we bother? There are a number of reasons for it, some of them are listed below:

1) As long as one is interested in *equilibrium* properties (and not in relaxational dynamics, aging, etc.) an **exact** ground state is more valuable than a non-equilibrium low temperature simulation.
2) One might expect that some features of the ground state persist at low temperatures (like the domain structure in the three-dimensional random field Ising model [fractal or not?], etc.).

3) If the phase transition into an ordered, maybe glassy, state happens at $T = 0$, one can extract critical exponents from ground state calculations (as for instance in the two-dimensional Ising spin glass).
4) If the RG (renormalization group) flow for a finite T transition is governed by a zero-temperature fixed point, one can again extract the critical exponents via ground state calculations (like in the three-dimensional random field Ising model). These are then, *if* the RG-picture is correct, identical with those for the finite T transition.
5) The zero temperature extrapolation of analytical finite T predictions for the glassy phase can be checked explicitly, like in the SOS (solid-on-solid) model with a disordered substrate.

As a motivation this should be enough, in the next section we jump directly *in medias res*. But before we start: many people think that combinatorial optimization is essentially the traveling salesman problem, only because it is by far the most famous problem (see [3] for an excellent introduction). This is similar to saying that frustrated disordered systems are essentially spin glasses (actually the traveling salesman problem and the spin glass problem are intimately connected via their complexity). One aim of this lecture is to remove this prejudice and to demonstrate that there are many more problems out there than only spin glasses (or traveling salesmen): algorithmically much easier to handle but equally fascinating. This is also the reason why on the algorithmic side we focus mainly on network flows: with the help of the material presented in the appendix everybody should be able to sit down in front of the computer and to implement efficiently the algorithms discussed there. If someone wants to know it *all*, i.e. all background material on graph theory, linear programming and network flows, we refer to standard works such as [4], [5], [6], [7], [8], [10].

4 Ground State Interface in a Random Medium

Although it was historically not the first random Ising model to be investigated with the help of the maximum flow/minimum cut algorithm (this was the random field Ising model, which we shall discuss later), it might be pedagogically more advantageous to start with the random bond Ising model with a boundary induced interface. The reason for the greater intuitive appeal of the latter problem is that the minimum cut, which the algorithm searches, is identical with the minimum energy interface of the physical system, which we seek.

The random bond Ising ferromagnet (RBIFM) is defined by

$$H = -\sum_{(ij)} J_{ij}\sigma_i\sigma_j \qquad (3)$$

with $\sigma_i = \pm 1$ Ising spins and $J_{ij} \geq 0$ ferromagnetic interactions strengths between neighboring spins. These are random quenched variables, which means that they are distributed according to some probability distribution and fixed right from the beginning. (ij) are nearest neighbor pairs on a $d+1$-dimensional lattice of size $L^d \times H$. We denote the coordinates by $i = (x_1, \ldots, x_d, y)$.

Since the interactions are all ferromagnetic, the ground state is simply given by $\sigma_i = +1$ for all sites i or $\sigma_i = -1$ for all sites i. Thus, up to here there is disorder, but no frustration in the problem. This changes by the boundary conditions (b.c.) we define now: we apply periodic b.c. in the x-directions and fix the spins at $y = 1$ to be $+1$ and those at $y = h$ to be -1.

$$\sigma_{(x_1,\ldots,x_d,y=1)} = +1 \quad \text{and} \quad \sigma_{(x_1,\ldots,x_d,y=h)} = -1 \qquad (4)$$

This induces an interface through the sample where bonds have to be broken, as indicated in Fig. 1. If all bonds were of the same strength $J_{ij} = J$ we would have the pure Ising model and the interface would simply be a d-dimensional hyperplane perpendicular to the y-direction, which costs an energy of JL^d, J for each broken bond. Because of the randomness of the J_{ij} it is energetically more favorable to break weak bonds: the interface becomes distorted and its shape is rough. This model has also been used to describe fractures in materials where the J_{ij} represents the local force needed to break the material and it is assumed that the fracture occurs along the surface of minimum total rupture force.

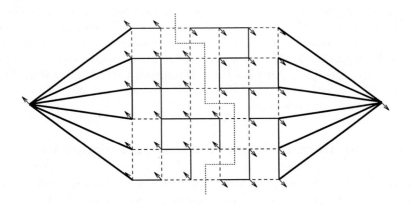

Fig. 1. Two-dimensional random bond Ising model with a binary distribution of interaction strengths $J_{ij} \in \{J_i, J_2\}$ with $J_1 \gg J_2 > 0$: Thick lattice bonds are strong ($J_{ij} = J_1$), broken lattice bonds are weak $J_{ij} = J_2$. The left and right boundary spins are connected to two extra spins $\sigma_s = +1$ and $\sigma_t = -1$, respectively, with infinitely strong bonds. ↖ means $\sigma_i = +1$, ↘ means $\sigma_i = -1$. The dotted line is the resulting interface: in this example it passes only through weak bonds, as a result of which it is of minimal energy. It partitions the lattice into up-and down-spins, the bonds (or more precisely: their corresponding forward arcs — see appendix A.3) that intersect the dotted line constitute the set $(S, \overline{S})$ for the s-t-cut $[S, \overline{S}]$.

How do we solve the task of finding the minimal energy configuration for the interface? First we map it onto a flow problem in a capacitated network (see appendix A for the nomenclature). We introduce two extra sites, a source node s, which we connect to all spins of the hyperplane $y = 1$ with bonds $J_{s,(x_1,...,x_d,y=1)} = J_\infty$, and a sink node t, which we connect to all spins of the hyperplane $y = h$ with bonds $J_{s,(x_1,...,x_d,y=h)} = J_\infty$. We choose $J_\infty = 2\sum_{(ij)} J_{ij}$, i.e. strong enough that the interface cannot pass through a bond involving one of the two extra sites. Now we enforce the b.c. (4) by simply fixing $\sigma_s = +1$ and $\sigma_t = -1$. The graph underlying the capacitated network we have to consider is now defined by the set of vertices (or nodes)

$$N = \{1, \ldots, L^{d+1}\} \cup \{s, t\} \tag{5}$$

and the set of edges (or arcs) connecting them

$$A = \{(i,j) | i, j \in N, J_{ij} > 0\}. \tag{6}$$

Note that we have forward *and* backward arcs for each pair of interacting sites in the lattice. The capacities u_{ij} of the arcs (i, j) are given by the bond strength J_{ij}. For any spin configuration $\sigma = (\sigma_1, \ldots, \sigma_N)$ we now define

$$S = \{i \in N | \sigma_i = +1\} \tag{7}$$
$$\overline{S} = \{i \in N | \sigma_i = -1\} = N \backslash S$$

Obviously $\sigma_s \in S$ and $\sigma_t \in \overline{S}$. Knowledge of S is sufficient to determine the energy of any spin configuration via (3):

$$H(S) = -\sum_{(i,j) \in E(S)} J_{ij} - \sum_{(i,j) \in E(\overline{S})} J_{ij} + \sum_{(i,j) \in (S,\overline{S})} J_{ij} \tag{8}$$
$$= -C + 2\sum_{(i,j) \in (S,\overline{S})} J_{ij}$$

where $E(S) = \{(i,j) | i \in S, j \in S\}$, $E(\overline{S}) = \{(i,j) | i \in \overline{S}, j \in \overline{S}\}$ and $(S,\overline{S}) = \{(i,j) | i \in S, j \in \overline{S}\}$. The constant $C = \sum_{(i,j) \in E(N)} J_{ij}$ is irrelevant (i.e. independent of S). Note that $(S, \overline{S})$ is the set of edges (or arcs) connecting S with $\overline{S}$, this means it **cuts** N in two disjoint sets. Since $s \in S$ and $t \in \overline{S}$, this is a so called *s-t-cut-set*, abbreviated $[S, \overline{S}]$. Thus the problem of finding the ground state of (3) with the interface inducing b.c. (4) can be reformulated as a *minimum cut* problem

$$\min_{S \subset N} \{H'(S)\} = \min_{[S,\overline{S}]} \sum_{(i,j) \in (S,\overline{S})} J_{ij}. \tag{9}$$

in the capacitated network (with $H' = (H + C)/2$) defined above. It does not come as a surprise that this minimum cut is *identical* with the interface between the $(\sigma_i=+1)$-domain and the $(\sigma_i=-1)$-domain that has the lowest energy. Actually any s-t-cut-set defines such an interface, some of them might consist of many components, which is of course energetically unfavorable.

To conclude, we have to find the minimum cut in a capacitated network, which is, as we show in appendix A, equivalent to finding a maximum flow from node s to node t. An intuitive argument for this famous max-flow-min-cut theorem is the following: Suppose you have to push, let's say waterflow through a network of pipelines, each with some capacity. The capacities in our case are the ferromagnetic interaction strengths on the bonds (pipes) between the nodes. Somewhere in the network there is a bottle-neck (in general consisting of several pipes) which does not allow a further increase of the waterflow sent from the source to the sink. If the maximum possible flow goes through the network, the flow on the pipes of the bottle-neck is equal to their capacity. The minimum cut is simply the global bottle-neck with the smallest capacity, and thus determines the maximum flow.

One can solve the above task by applying the straightforward augmenting path algorithm discussed in appendix A.2, which is based on the idea of finding directed paths in the network along which one could possibly send more flow from the source to the sink. If one finds such paths, one augments the flow along them (i.e. pushes more water through the pipes), if there are none, the present flow is optimal. In the latter case one identifies the corresponding s-t cut, which then yields the exact ground state interface for the above problem.

A more efficient way is to use the preflow push algorithm presented in appendix A.4. The idea of this algorithm is to *flood* the network starting from the source. Then one encounters the situation that some nodes are not able to transport the flood coming from the source in the direction of the sink, which means that one has to send some flow back to the source. The time consuming part of this algorithm is the retreat of floods that have been pushed too far, and this retreat happens faster if the capacities of the backwards arcs are as large as possible. Bearing this observation in mind Middleton [11] has suggested a nice modification of the original problem that yields a significant speed up: to forbid *overhangs* of the interface we are discussing is equivalent to introducing backward arcs with infinite capacity in the corresponding flow network (obviously a minimum cut will then never contain such an arc as forward arc). Thus, in the event that too much flow has been pushed, the retreat works with maximum efficiency.

To conclude let us cite a number of results that have been obtained in this way. Of particular interest here is the width of the interface

$$W(L,h) = ([y_x^2]_{\text{av}} - [y_x]_{\text{av}}^2)^{1/2} \sim L^\zeta \tilde{w}(h/L^\zeta) \,, \tag{10}$$

where y_x is the y-coordinate of the point (x,y) of the interface (note that because of the "no overhangs" prescription y_x is single-valued). $[\ldots]_{\text{av}}$ means an average over the disorder. One expects the finite size scaling form as indicated with a roughness exponent ζ. From the ground state calculations and finite size scaling one finds [11] that $\zeta = 0.41 \pm 0.01$ in 2d (L up to 120, and h up to 50, with $10^3 - 10^4$ samples), and $\zeta = 0.22 \pm 0.01$ in 3d (L up to 30 and h up to 20).

5 The Random Field Ising Model

The random field Ising model (RFIM, for a review see [12]) is defined

$$H = -\sum_{(ij)} J_{ij}\sigma_i\sigma_j - \sum_i h_i\sigma_i \quad (11)$$

with $\sigma_i = \pm 1$ Ising spins, ferromagnetic bonds $J_{ij} \geq 0$ (random or uniform), (ij) nearest neighbor pairs on a d-dimensional lattice and at each site i a random field $h_i \in R$ that can be positive or negative. The first term prefers a ferromagnetic order, which means it tries to align all spins. The random field, however, tends to align the spins with the field which points in random directions depending on whether it is positive or negative. This is the source of frustration in this model.

Let us suppose for the moment uniform interactions $J_{ij} = J$ and a symmetric distribution of the random fields with mean zero and variance h_r. It is established by now that in 3d (and higher dimensions) the RFIM shows ferromagnetic long range order at low temperatures, provided h_r is small enough. In 1d and 2d there is no ordered phase at any finite temperature. Thus in 3d one has a paramagnetic/ferromagnetic phase transition along a line $h_c(T)$ in the h_r-T-diagram.

The renormalization group (RG) picture says that the nature of the transition is the same[2] all along the line $h_c(T)$, with the exception of the pure fixed point at $h_r = 0$ and $T_c \sim 4.51J$. The RG flow is dominated by a zero temperature fixed point at $h_c(T{=}0)$. As a consequence, the critical exponents determining the critical behavior of the RFIM should all be identical along the phase transition line, in particular identical to those one obtains *at zero temperature* by varying h_r alone.

Therefore we consider zero temperature from now on. Close to the transition at $h_c = h_c(T{=}0)$ one would e.g. expect for the disconnected susceptibility

$$\chi_{\text{dis}} = \frac{1}{L^3}\left[\sum_{i,j}\sigma_i\sigma_j\right]_{\text{av}} \sim L^{4-\bar{\eta}}\tilde{\chi}(L^{1/\nu}\delta) \quad (12)$$

where $\delta = h_r - h_c$ is the distance from the critical point and ν is the thermal critical exponent. An analogous expression holds for the magnetization involving the exponent β. Thus to estimate a set of critical exponents the task is to calculate the ground state configurations of the RFIM (11).

This optimization task is again equivalent to a maximum flow problem [13], as in the interface model discussed in the last section. Historically the RFIM was the first physical model investigated with a maximum flow algorithm [14]. However, here the minimum-cut is not a geometric object within the original system and therefore we found it more intuitive to discuss the RFIM after the interface problem.

---

[2] We leave aside the discussion about a possible tricritical point (which does not seem to be the case [12]) and the existence of an intervening spin glass phase.

In essence we proceed in the same way as in the last section. Again we add to extra nodes s and t and put spins with fixed values there:

$$\sigma_s = +1 \quad \text{and} \quad \sigma_t = -1 \tag{13}$$

We connect all sites with positive random field to the node s and all sites with negative random field to t:

$$J_{si} = \begin{cases} h_i & \text{if } h_i \geq 0 \\ 0 & \text{if } h_i < 0 \end{cases}$$
$$J_{it} = \begin{cases} |h_i| & \text{if } h_i < 0 \\ 0 & \text{if } h_i \geq 0 \end{cases} \tag{14}$$

We construct a network with the set of nodes $N = \{1, \cdots, L^d\} \cup \{s, t\}$ and the set of (forward and backward) arcs $A = \{(i,j)| i, j \in N, J_{ij} > 0\}$. Each of them has a capacity $u_{ij} = J_{ij}$. Now we can write the energy or cost function as

$$E = -\sum_{(i,j) \in A} J_{ij} \sigma_i \sigma_j \tag{15}$$

and, by denoting the set $S = \{i \in N | S_i = +1\}$ and $\overline{S} = N \backslash S$ the energy can be written as in equation (8):

$$E(S) = -C + 2 \sum_{(i,j) \in (S, \overline{S})} J_{ij} \tag{16}$$

with $C = \sum_{(i,j) \in A} J_{ij}$. The problem is reduced to the problem of finding a minimum s-t-cut as in (9). The difference from the interface problem is that now the extra bonds connecting the two special nodes s and t with the original lattice do not have infinite capacity: they can lie *in* the cut, namely whenever it is more favorable not to break a ferromagnetic bond but to disalign a spin with its local random field. In the extended graph which we consider the s-t-cut again forms a connected interface, however, in the original lattice (without the bonds leading to and from the extra nodes) the resulting structure is generally *disconnected*, a multicomponent interface. Each single component surrounds a connected region in the original lattice containing spins, which all point in the same direction. In other words, they form ferromagnetically ordered domains separated by domain walls given by the subset of the s-t-cut that lies in the original lattice.

The maximum flow algorithm has been used by Ogielski [14] to calculate the critical exponents of the RFIM via the above mentioned finite size scaling. He obtained

$$\nu = 1.0 \pm 0.1, \quad \overline{\eta} = 1.1 \pm 0.2, \quad \beta \approx 0.05 \tag{17}$$

with β being so small that it is (numerically) indistinguishable from zero, indicating a *discontinuous* transition. These estimates are compatible with those obtained by recent Monte Carlo simulations supporting the RG idea of the universality of the transition at finite *and* zero temperature. However, this is still not the end of the story, since various scaling predictions, also based on the RG picture, are violated. For further details we refer to the review [12].

6 The Diluted Antiferromagnet in an External Field

Experimentally it is of course hard to prepare a random field at each lattice site, therefore one might ask why people have been so enthusiastic about the RFIM, discussed in the last section, over decades. Actually it is because within a field theoretic perturbation theory (around small random fields) it has been shown [15] that the RFIM is in the same universality class as the diluted antiferromagnet in a *uniform* magnetic field (DAFF) defined via

$$H = +\sum_{(ij)} J_{ij}\varepsilon_i\varepsilon_j\sigma_i\sigma_j - \sum_i h_i\varepsilon_i\sigma_i \tag{18}$$

where $\sigma_i = \pm 1$, $J_{ij} \geq 0$, (ij) are nearest neighbor pairs on a simple cubic lattice, and $\varepsilon_i \in \{0,1\}$ with $\varepsilon_i = 1$ with probability c, representing the concentration of spins. Usually one takes $J_{ij} = J$ and $h_i = h$, both uniform. Because of the plus sign in front of the first term in (18) all interactions are antiferromagnetic, the model represents a diluted antiferromagnet, for which many experimental realizations exist (e.g. $Fe_cZn_{1-c}F_2$). Now that neighboring spins tend to point in opposite directions due to their antiferromagnetic interaction a uniform field competes with this ordering tendency by trying to align them all. Thus it is again a frustrated system. Due to the analogy with the RFIM model one expects at low temperatures and small enough fields a second order phase transition from a paramagnetic to an antiferromagnetic phase.

In recent years people began to doubt the folklore that the DAFF is under all circumstances a good experimental realization of the RFIM model. Note that this result has been derived for small fields h, and the question is whether this still holds at larger fields. The largest field value at which the paramagnetic-antiferromagnetic transition can be studied is $h_c(T = 0)$. This motivates the study of the ground state transition along the same lines as in the RFIM context. Preliminary results [17] indicate that the critical exponents are *different* here, which implies that the RFIM and the DAFF are in different universality classes at large field values.

Here we are primarily interested in the question of whether we can again map the calculation of ground states onto a maximum flow problem, as for the RFIM. The answer is yes as long as the antiferromagnetic interactions are short ranged among nearest neighbors on a bipartite lattice. With zero external field the ground state would be antiferromagnetic, which means (remember we have a simple cubic lattice) that we can define two bipartite sublattices A and B like the black and white fields of a checkerboard. Each site i in A finds all its nearest neighbors j in B and vice versa. Define new spin and field variables via

$$\sigma'_i = \begin{cases} +\sigma_i & \text{for} \quad i \in A \\ -\sigma_i & \text{for} \quad i \in B \end{cases} \quad , \quad h'_i = \begin{cases} +\varepsilon_i h_i & \text{for} \quad i \in A \\ -\varepsilon_i h_i & \text{for} \quad i \in B \end{cases} . \tag{19}$$

Since $\sigma'_i\sigma'_j = -\sigma_i\sigma_j$ for all nearest neighbor pairs (ij) one can write (18) as

$$H = -\sum_{(ij)} J'_{ij}\sigma'_i\sigma'_j - \sum_i h'_i\sigma'_i \tag{20}$$

with $J'_{ij} = J_{ij}\varepsilon_i\varepsilon_j$. Now the Hamiltonian has exactly the same form as the one for the RFIM, since $J'_{ij} \geq 0$. Note that even if one starts with uniform bonds $J_{ij} = J$ and a uniform field $h_i = h$ the dilution generates bond- and field disorder! Now that one has reduced the problem to the RFIM we also know how to map it to a maximum flow problem. Hartmann and Usadel [16] have extended the algorithm in such a way that *all* ground states can be calculated: for uniform bonds and fields the resulting RFIM has a discrete distribution of random bonds and fields, which leads in general to a high degeneracy of the ground state, something that does not happen in the case of a continuous distribution, where usually the ground state is unique.

In this context we would like to mention the Coulomb glass model [18], [19], which is a model for point charges on a d-dimensional lattice with long-range Coulomb interactions (repulsive of strength V/r with V *positive* and r being the Euclidean distance between two charges):

$$H = \sum_{i,j} \frac{V}{r_{ij}} n_i n_j + \sum_i n_i \mu_i \qquad (21)$$

where now the sum is over *all* pairs of sites of the lattice. $n_i \in \{0, 1\}$ indicates the presence ($n_i = 1$) or absence ($n_i = 0$) of a charged particle at site i and r_{ij} is the Euclidean distance between site i and site j. The random local potentials $\mu_i \subset [-W, W]$ represent the quenched disorder. Obviously this model is equivalent to an antiferromagnet with long-range interactions and random fields. As yet now no way of mapping this interesting problem onto a network flow problem is known, it seems to fall into the spin glass class, which we discuss now.

7 The Spin Glass Problem

Spin glasses are the prototypes of (disordered) frustrated systems, their history is quite a long one and for the present status of numerical investigation I refer to [12], where numerous references to experimental and theoretical introductions can also be found. In the models we have discussed up to now, the frustration was caused by two separate terms of different physical origin (interactions and external fields or boundary conditions). Spin glasses are magnetic systems in which the magnetic moments interact ferro- or antiferromagnetically in a random way, as in the following Edwards-Anderson Hamiltonian for a short ranged Ising spin glass (SG)

$$H = -\sum_{(ij)} J_{ij}\sigma_i\sigma_j , \qquad (22)$$

where $\sigma_i = \pm 1$, (ij) are nearest neighbor sites on a d-dimensional lattice and the interaction strengths $J_{ij} \in R$ are unrestricted in sign. By analogy to (8–9) one shows that the problem of finding the ground state is again equivalent to finding a minimal cut $[S, \overline{S}]$ in a network

$$\min_{\sigma}\{H'(\underline{\sigma})\} = \min_{[S,\overline{S}]} \sum_{(i,j)\in(S,\overline{S})} J_{ij} , \qquad (23)$$

again $H' = (H+C)/2$ with $C = \sum_{(ij)} J_{ij}$. However, now the capacities $u_{ij} = J_{ij}$ of the underlying network are *not* non-negative any more, therefore it is *not* a minimum-cut problem in the sense of appendix A.3 and thus it is also not equivalent to a maximum flow problem, which we know how to handle efficiently.

It turns out that the spin glass problem is *much* harder than the questions we have discussed so far. In general (i.e. in any dimension larger than two and also for 2d in the presence of an external field) the problem of finding the SG ground state is $\mathcal{NP}$-complete [20], which means in essence that no polynomial algorithm for it is known (and also that the chances of finding one in the future are marginal). Nevertheless, some extremely efficient algorithms for it have been developed [22], [23], [24], which have a non-polynomial bound for their worst case running-time but which terminate (i.e find the optimal solution) after a reasonable computing time for pretty respectable system sizes.

First we discuss the only non-trivial case that can be solved with a polynomial algorithm: the two-dimensional Ising SG on a planar graph. This problem can be shown to be equivalent to finding a minimum weight perfect matching, which can be solved in polynomial time. We do not treat matching problems and the algorithms to solve them in this lecture (see [5], [6], [8]), however, we would like to present the idea [20]. To be concrete let us consider a square lattice with free boundary conditions. Given a spin configuration $\underline{\sigma}$ (which is equivalent to $-\underline{\sigma}$) we say that an edge (or arc) (i, j) is satisfied if $J_{ij}\sigma_i\sigma_j > 0$ and it is *unsatisfied* if $J_{ij}\sigma_i\sigma_j < 0$. Furthermore we say a plaquette (i.e. a unit cell of the square lattice) is *frustrated* if it is surrounded by an odd number of negative bonds (i.e. $J_{ij} \cdot J_{jk} \cdot J_{kl} \cdot J_{li} < 0$ with i, j, k and l the four corners of the plaquette)). There is a one-to-one correspondence between equivalent spin configurations ($\underline{\sigma}$ and $-\underline{\sigma}$) and sets of unsatisfied edges with the property that on each frustrated (unfrustrated) plaquette there is an odd (even) number of unsatisfied edges. See Fig. 2 for illustration.

Note that

$$H(\underline{\sigma}) = -C + 2 \sum_{\text{unsatisfied edges}} |J_{ij}| . \qquad (24)$$

which means that one has to minimize the sum over the weights of unsatisfied edges. A set of unsatisfied edges will be constituted by a set of paths (in the dual lattice) from one frustrated plaquette to another and a set of closed circles (see Fig. 2). Obviously the latter always increase the energy so that we can neglect them. The problem of finding the ground state is therefore equivalent to finding the minimum possible sum of the weights of these paths between the frustrated plaquettes. This means that we have to *match* the black dots in the Fig. 2 with one another in an optimal way. One can map this problem on a minimum weight *perfect matching*[3] problem, which can be solved in polynomial time (see [20] for further details).

---
[3] A perfect matching of a graph $G = (N, A)$ is a set $M \subseteq A$ such that each node has only one edge of M adjacent to it.

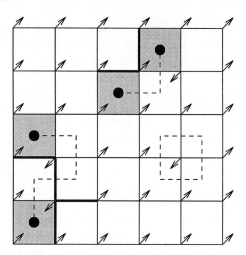

Fig. 2. Two-dimensional Ising spin glass with $\pm$-J couplings: Thin lines, are positive interactions, thick lines are negative interactions, ↗ means $\sigma_i = +1$, ↙ means $\sigma_i = -1$, shaded faces are frustrated plaquettes, broken lines cross unsatisfied edges.

Note that for binary couplings, i.e. $J_{ij} = \pm J$, where $J_{ij} = +J$ with probability p the weight of a matching is simply proportional to the sum of the lengths of the various paths connecting the centers of the frustrated plaquettes, which simplifies the actual implementation of the algorithm. In [21] the 2d $\pm J$ spin glass and the site disordered SG[4] has been studied extensively with this algorithm.

As we mentioned, in any case other than except the planar lattice situation discussed above the spin glass problem is $\mathcal{NP}$-hard. In what follows we would like to sketch the idea of an efficient but non-polynomial algorithm [23], [25]. To avoid confusion with the minimum cut problem we discussed in connection with maximum flows one calls the problem (23) a max-cut problem (since finding the minimum of H is equivalent to finding the maximum of $-H$).

Let us consider the vector space R^A. For each cut $[S, \overline{S}]$ define $\chi^{(S,\overline{S})} \in R^A$, the incidence vector of the cut, by $\chi_e^{(S,\overline{S})} = 1$ for each edge $e = (i,j) \in (S, \overline{S})$ and $\chi_e^{(S,\overline{S})} = 0$ otherwise. Thus there is a one-to-one correspondence between cuts in G and their $\{0, 1\}$-incidence vectors in R^A. The *cut-polytope* $P_C(G)$ of G is the convex hull of all incidence vectors of cuts in G: $P_C(G) = \text{conv}\,\{\chi^{(S,\overline{S})} \in R^A \,|\, S \subseteq A\}$. Then the max-cut problem can be written as a *linear program*

$$\max\,\{\underline{u}^T \underline{x} \,|\, \underline{x} \in P_C(G)\} \tag{25}$$

---

[4] The site disordered spin glass is defined as follows: occupy the sites of a square lattice randomly with A (with concentration c) and B (with concentration $1 - c$) atoms. Now define the interactions J_{ij} between neighboring atoms: $J_{ij} = -J$ if there are A-atoms on both sites and $J_{ij} = +J$ otherwise.

since the vertices of $P_C(G)$ are cuts of G and vice versa. Linear programs usually consist of a linear cost function $\underline{u}^T\underline{x}$ that has to be maximized under the constraint of various inequalities defining a Polytope in R^n (i.e. the convex hull of finite subsets of R^n) and can be solved for example by the simplex method, which proceeds from corner to corner of that polytop to find the maximum (see e.g. [5], [7], [8]). The crucial problem in the present case is that it is $\mathcal{NP}$-hard to write down all inequalities that represent the cut polytop $P_C(G)$.

It turns out that *partial* systems are also useful, and this is the essential idea for an efficient algorithm to solve the general spin glass problem as well as the traveling salesman problem or other so called mixed integer problems (i.e. linear programs where some of the variables x are only allowed to take on some integer values, like 0 and 1 in our case) [3], [26]. One chooses a system of linear inequalities L whose solution set $P(L)$ contains $P_C(G)$ and for which $P_C(G)$ = convex hull $\{\mathbf{x} \in P(L) | x \text{ integer}\}$. In the present case these are $0 \leq x \leq 1$, which is trivial, and the so called cycle inequalities, which are based on the observation that all cycles in G have to intersect a cut an even number of times (have a look at the cut in Fig. 1 and choose as cycles for instance the paths around elementary plaquettes). The most remarkable feature of this set L of inequalities is that the separation problem[5] for them can be solved in polynomial time: the **cutting plane algorithm** which, starting from some small initial set of inequalities, generates iteratively new inequalities until the optimal solution for the actual subset of inequalities, is feasible. Note that one does not solve this linear programm by the simplex method since the cycle inequalities are still too numerous for this to work efficiently.

Due to the insufficient knowledge of the inequalities that are necessary to describe $P_C(G)$ completely, one may end up with a nonintegral solution $\mathbf{x}^*$. In this case one **branches** on some fractional variable x_e (i.e. a variable with $x_e^* \notin \{0, 1\}$), creating two subproblems in one of which x_e is set to 0 and in the other x_e is set to 1. Then one applies the cutting plane algorithm recursively for both subproblems, which is the origin of the name **branch-and-cut**. Note that in principle this algorithm is not restricted to any dimension, boundary conditions, or to the fieldless case. However, there are realizations of it that run fast (e.g. in 2d) and others that run slow (e.g. in 3d) and it is an ongoing research aim to improve on the latter. A detailed description of the rather complex algorithm can be found in [26], [25].

The typical question one tries to address in the context of spin glasses is: is there a spin glass transition at finite temperature, below which the spins freeze into some configuration (i.e. $\langle \sigma_i \rangle_T \neq 0$ for $T < T_c$)? What can we do with ground state calculation to answer this question? Here the concept of the domain wall energy plays a crucial role [27]. What a finite but small temperature does is to

---

[5] The seperation problem for a set of inequalities L consists in either proving that a vector x satisfies all inequalities of this class or in finding an inequality that is violated by $\mathbf{x}$. A linear program can be solved in polynomial time if and only if the separation problem is solvable in polynomial time [9].

destroy the ground state order by collectively flipping larger and larger clusters (droplets). If the energy cost for a reversal of a cluster of linear size L increases with L (like $\Delta E \propto L^y$ with $y > 0$) thermal fluctuation will not be able to destroy long range order, and thus we have a spin glass transition at finite T_c. If it decreases (i.e. $y < 0$) long range order is unstable with respect to thermal fluctuations and the spin glass state occurs only at $T = 0$. As an example consider the d-dimensional *pure* Ising ferromagnet, for which the ground state is all spins up or all down. Reversing a cluster of linear size L breaks all surface bonds of this cluster, which means that it costs an energy $\Delta E \propto L^{d-1}$, i.e. $y = d - 1$ for the pure ferromagnet. Thus the ferromagnetic state in pure Ising systems is stable for $d > 1$, which is well known. Instead of reversing spins one usually studies the energy difference between ground states for periodic and antiperiodic boundary conditions. In [28] it has been shown that

$$\Delta E \sim L^y \tag{26}$$

with $y = -0.281$ for the 2d Ising spin glass with a uniform distribution (thus there is no finite T SG transition in this case). It has been speculated that in the $\pm J$ case for a range of concentration of ferromagnetic bonds [29] and in the site-random case for some concentration of A atoms [30] a spin glass phase might exist at non-zero temperature $T > 0$. This possibility has been ruled out in [21] with the help of ground state calculations.

With the above mentioned branch & cut algorithm the magnetic field dependence of the ground state magnetization $m_L(h) = L^{-d}[|\sum_i \sigma_i|]_{\mathrm{av}}$ has been calculated in the 2d case with a continuous coupling distribution. In [28] it has been shown that it obeys finite size scaling:

$$m_L(h) \sim L^{-d/2} \tilde{m}(Lh^{1/\delta}) \tag{27}$$

(note $d = 2$) with $\delta = 1.481$. This value is remarkable in so far as it clearly violates the scaling prediction $\delta = 1 - y$.

Finally we would like to focus our attention on the very important concept of *chaos* in spin glasses. This notion implies an extreme sensitivity of the SG-state with respect to small parameter changes like temperature or field variations. There is a length scale λ — the so called overlap length — beyond which the spin configurations within the same sample become completely decorrelated if compared for instance at two different temperatures

$$C_{\Delta T} := [\langle \sigma_i \sigma_{i+r} \rangle_T \langle \sigma_i \sigma_{i+r} \rangle_{T+\Delta T}]_{\mathrm{av}} \sim \exp\left(-r/\lambda(\Delta T)\right). \tag{28}$$

This should also hold for the ground states if one slightly varies the interaction strengths J_{ij} in a random manner with amplitude δ. Let $\underline{\sigma}$ be the ground state of a sample with couplings J_{ij} and $\underline{\sigma}'$ the ground state of a sample with couplings $J_{ij} + \delta K_{ij}$, where the K_{ij} are random (with zero mean and variance one) and δ is a small amplitude. Now define the overlap correlation function as

$$C_\delta(r) := [\sigma_i \sigma_{i+r} \sigma'_i \sigma'_{i+r}]_{\mathrm{av}} \sim \tilde{c}(r\delta^{1/\zeta}), \tag{29}$$

where the last relation indicates the scaling behavior we would expect (the overlap length being $\lambda \sim \delta^{-1/\zeta}$) and ζ is the *chaos* exponent. In [28] this scaling prediction was confirmed with $1/\zeta = 1.2 \pm 0.1$.

8 The SOS-Model on a Disordered Substrate

Up to now we have considered Ising models exclusively. Quite recently it has been shown [31], [32] that many more frustrated systems are amenable to ground state studies of the kind we have discussed so far. Consider a solid-on-solid model with random offsets, modeling a crystalline surface on a disordered substrate as indicated in Fig. 3. It is defined by the following Hamiltonian (or energy function):

$$H = \sum_{(ij)} f(h_i - h_j) \qquad (30)$$

where (ij) are nearest neighbor pairs on a d–dimensional lattice ($d = 1, 2$) and $f(x)$ is an arbitrary convex ($f''(x) \geq 0$) and symmetric ($f(x) = f(-x)$) function, for instance $f(x) = x^2$. Each height variable $h_i = d_i + n_i$ is the sum of an integer particle number which can also be negative, and a substrate offset $d_i \in [0, 1[$. For a flat substrate, $d_i = 0$ for all sites i, we have the well known SOS-model [34]. The disordered substrate is modeled by random offsets $d_i \in [0, 1[$ [33], which are distributed independently.

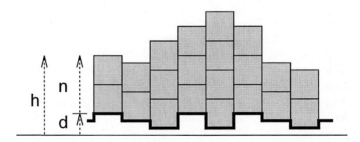

Fig. 3. The SOS model on a disordered substrate. The substrate heights are denoted by $d_i \in [0, 1]$, the number of particle on site i by $n_i \in Z$, which means that they could also be negative, and the total height on site i by $h_i = d_i + n_i$

The model (30) has a phase transition at a temperature T_c from a (thermally) rough phase for $T > T_c$ to a *super-rough* low temperature phase for $T < T_c$. In two dimensions "rough" means that the height-height correlation function diverges logarithmically with the distance $C(r) = [\langle (h_i - h_{i+r})^2 \rangle]_{\rm av} = a \cdot T \cdot \log(r)$ (with $a = 1/\pi$ for $f(x) = x^2$), "super-rough" means that either the prefactor in front of the logarithm is significantly larger than $a \cdot T$, or that $C(r)$ diverges faster than $\log(r)$, e.g. $C(r) \propto \log^2(r)$.

Part of the motivation to study this model thus comes from its relation to flux lines in disordered superconductors, in particular high-T_c superconductors: The phase transition occurring for (30) is in the same universality class as a flux line array with point disorder defined via the two-dimensional Sine-Gordon model with random phase shifts

$$H = -\sum_{(ij)}(\phi_i - \phi_j)^2 - \lambda \sum_i \cos(\phi_i - \theta_i), \tag{31}$$

where $\phi_i \in [0, 2\pi[$ are phase variables, $\theta_i \in [0, 2\pi[$ are quenched random phase shifts and λ is a coupling constant. One might anticipate that both models (30) and (31) are closely related by realizing that both have the same symmetries (the energy is invariant under the replacement $n_i \to n_i + m$ ($\phi_i \to \phi_i + 2\pi m$) with m an integer). Close to the transition one can show that all higher order harmonics apart from the one present in the Sine-Gordon model (31) are irrelevant in a field theory for (30), which establishes the identity of the universality classes[6].

To calculate the ground states of the SOS model on a disordered substrate with general interaction function $f(x)$ we map it onto a minimum cost flow model. Let us remark, however, that the special case $f(x) = |x|$ can be mapped onto the interface problem in the random bond Ising ferromagnet in 3d with *columnar* disorder [35] (i.e. all bonds in a particular direction are identical), by which it can be treated with the maximum flow algorithm we know already.

We define a network G by the set of nodes N being the sites of the dual lattice of our original problem and the set of *directed* arcs A connecting nearest neighbor sites (in the dual lattice) (i,j) and (j,i). If we have a set of height variables n_i we define a flow **x** in the following way: Suppose two neighboring sites i and j have a positive (!) height difference $n_i - n_j > 0$. Then we assign the flow value $x_{ij} = n_i - n_j$ to the directed arc (i,j) in the dual lattice, for which the site i with the larger height value is on the right hand side, and assign zero to the opposite arc (j,i), i.e. $x_{ji} = 0$. And also $x_{ij} = 0$ whenever site i and j are of the same height. See Fig. 4 for a visualization of this scheme. Obviously then we have:

$$\forall i \in N : \sum_{\{j\,|\,(i,j)\in A\}} x_{ij} = \sum_{\{j\,|\,(j,i)\in A\}} x_{ji}. \tag{32}$$

On the other hand, for an arbitrary set of values for x_{ij} the constraint (32) has to be fulfilled in order to be a flow, i.e. in order to allow a reconstruction of height variables from the height differences. This observation becomes immediately clear by looking at Fig. 4.

We can rewrite the energy function as

$$H(\mathbf{x}) = \sum_{(i,j)} h_{ij}(x_{ij}), \quad \text{with} \quad c_{ij}(x) = f(x - d_{ij}), \tag{33}$$

---

[6] Note, however, that that far away from T_c, as for instance at zero temperature, there might be differences between the two models.

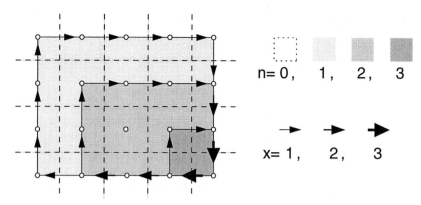

Fig. 4. The flow representation of a surface (here a "mountain" of height $n_i = 3$). The broken lines represent the original lattice, the open dots are the nodes of the dual lattice. The arrows indicate a flow on the dual lattice, that results from the height differences of the variables n_i on the original lattice. Thin arrows indicate a height difference of $x_{ij} = 1$, medium arrows $x_{ij} = 2$ and thick arrows $x_{ij} = 3$. According to our convention the larger height values are always on the *right* of an arrow. Observe that the mass balance constraint (32) is fulfilled on each node.

with $d_{ij} = d_i - d_j$. Thus our task is to minimize $H(\mathbf{x})$ under the constraint (32), which is (see appendix C.5) a *minimum-cost-flow* problem with the mass balance constraints (32) and arc *convex* cost functions $h_{ij}(x_{ij})$. It is worth mentioning that this mapping from the SOS model to a flow problem is closely related to the dual link representation of the XY-model in two dimensions [36]. This does not come as a great surprise since we already pointed out the relationship with the Sine-Gordon Hamiltonian involving phase variables (31).

The most straightforward way to solve this problem is to start with all height variables set to zero (i.e. $\mathbf{x} = 0$) and then to look for regions (or clusters) that can be increased collectively by one unit with a gain in energy. This is essentially what the negative cycle canceling algorithm discussed in appendix C.5 does: The negative cycles in the dual lattice surround the regions in which the height variables should be increased or decreased by one. However, it turns out that this is a non-polynomial algorithm, the so called successive shortest path algorithm is more efficient and solves this problem in polynomial time, see appendix C.5. This algorithm starts with an optimal solution for $H(\mathbf{x})$, which is given by $x_{ij} = +1$ for $d_{ij} > 1/2$, $x_{ij} = -1$ for $d_{ij} < -1/2$ and $x_{ij} = 0$ for $d_{ij} \in [-1/2, +1/2]$. Since this set of flow variables violates the mass balance constraints (32) (in general there is some imbalance at the nodes) the algorithm iteratively removes the excess/deficit at the nodes by augmenting flow.

Let us briefly summarize the results one obtains by applying this algorithm to the ground state problem for the SOS model on a disordered substrate [32]:

- The height-height correlation function diverges as $C(r) \propto \log^2(r)$ with the distance r.

- $\chi_L = L^{-4}\sum_{i,j}[(h_i - h_j)^2]_{\rm av}$ can nicely be fitted to $\chi_L = a + b\log(L) + c\log^2(L)$, indicating again a $\log^2$ dependence of the height-height correlation function. Moreover, the coefficients a, b and c depend on the power n in $f(x) = |x|^n$: c increases systematically with increasing n.
- By considering a boundary induced step in the ground state configuration one sees that the step energy increases algebraically with the system size: $\Delta E \sim L^y$ with $y = 0.45 \pm 0.05$. This corresponds to the domain wall energy introduced in the context of spin glasses in the last section. Furthermore the step is *fractal* with a fractal dimension close to 3/2.
- Upon a small, random variation of the substrate heights d_i of amplitude δ the ground state configuration decorrelates beyond a length scale $\lambda \sim \delta^\eta$ with $\eta = 0.95 \pm 0.05$. This implies the *chaotic* nature of the glassy phase in this model in analogy to spin glasses.

We would like to mention that this mapping of the original SOS model (30) on the flow problem works only for a planar graph (i.e. free or fixed boundary conditions), otherwise it is not always possible to reconstruct the height variables n_i from the height differences x_{ij}. As a counterexample in a toroidal topology (periodic boundary conditions) consider a flow, which is zero everywhere except on a circle looping the torus, where it is one. Although this flow fulfills the mass balance constraints (which are local) it is globally inadmissible: To the right of this circle the heights should be one unit larger than on left, but left and right become interchanged by looping the torus in the perpendicular direction, which causes a contradiction. If one insists on periodic boundary conditions, which have some advantages due to the translational invariance, one should recur to the special case $f(x) = |x|$, which can be treated differently, as we mentioned at the beginning of this section.

9 Vortex Glasses and Traffic

Finally we would like to focus on some further applications of the minimum cost flow algorithms that we discussed in the last section. Since we deal with network flow problems it should not come as a surprise that a number of physical problems involving magnetic flux lines can be mapped onto them. We have already mentioned the Sine-Gordon model with random phase shifts (31) describing a flux line array with point disorder and which is related to the SOS model on a disordered substrate. This relationship can be made more concrete with the help of the triangular Ising SOS model as discussed in [35].

The gauge glass model describes the **vortex glass** transition in three-dimensional superconductors. If one includes the screening of the interactions between vortices one can show that in the strong screening limit, the model Hamiltonian (in the link representation) acquires the form [38]

$$H = \sum_{(i,j)} (x_{ij} - b_{ij})^2 \qquad (34)$$

where x_{ij} are integer vortex variables living on the links (i,j) of the dual of the original simple cubic lattice. They represent magnetic flux lines, therefore their divergence has to be zero – which means that they have to fulfill the mass balance constraint (32). The quenched random variables b_{ij} also fulfill the same constraint (they have to be constructed as a lattice curl from a quenched vector potential). Moreover one has periodic boundary condition.

It has been shown that this model has a vortex glass transition at zero temperature. Thus, for the characterization of the critical behavior either low temperature Monte Carlo simulations [38] or ground state calculations become mandatory. The latter program has been performed in a tentative way in [39] with a stochastic, non-exact method for small system sizes ($L \leq 4$). The problem of minimizing (34) under the above mentioned constraints is a convex cost flow problem that can be solved in a straightforward manner with the algorithms presented in appendix C.5. Work is in progress in this direction [37].

A further application of the minimum cost flow algorithms with convex cost functions is **traffic flow**, which became a major research topic in *physics* quite recently [40]. Network flow problems naturally occur in any transportation system: what is the shortest path between point A and point B in a road network (shortest path problem), how many vessels does a shipping company need to have in order to deliver perishable goods between several different origin–destination pairs (maximum flow problem) or what is the flow that satisfies the demands at a number of warehouses from the available supplies at a number of plants and that minimizes the shipping cost (typical transportation problem = minimum cost flow problem).

All of the above problems are *linear* problems. Whenever system congestion or queuing effects have to be taken into account in the model describing a "real" network flow, the introduction of nonlinear costs (since queuing delays vary nonlinearily with flows) are mandatory. In road networks, as more vehicles use any road segment, the road becomes increasingly congested and so the delay on that road increases. For example, the delay on a particular road segment, as a function of the flow x on that road, might be $\alpha x/(u-x)$. In this expression u denotes a theoretical capacity of the road and α is another constant: As the flow increases, so does the delay; moreover, as the flow x approaches the theoretical capacity of that road segment, the delay on the link becomes arbitrarily large. In many instances, as in this example, the delay function on each road segment is a *convex* function of the road segment's flow, so finding the flow plan that achieves the minimum overall delay, summed over all road segments, is a convex cost network flow model.

It should have become clear by the end of this lecture that frustrated, disordered systems and network flows are closely related, or even equivalent. The quenched disorder occurring in the physical models we have discussed so far find their counterpart in arc capacities and costs in flow problems. Thus, many "daily life" networks, like transportation systems or urban traffic flows, share many features with disordered or even glassy systems. For instance the concept of *chaos* we encountered in spin glasses as well as in the random solid-on-solid model should also be valid in traffic networks: the slightest random (i.e. uncontrollable) change in the capacities of the roads, as for instance after a heavy rain or snowfall, or locally by several accidents, should completely change the traffic flow pattern beyond a particular length scale. A systematic study of these issues is certainly of great interest, not only for the theoretical understanding of the intrinsically chaotic nature of complex network flows but also for practical reasons.

Acknowledgement: I would like to express my special thanks to M. Jünger, U. Blasum, M. Diehl and N. Kawashima, from whom I learnt about various aspects of the issues treated in this lecture and with whom I enjoy(ed) an ongoing fruitful and lively collaboration. This work has been supported by the Deutsche Forschungsgemeinschaft (DFG).

Appendix: Concepts in Network Flows and Basic Algorithms

In this appendix we introduce the basic definitions in the theory of network flows, which are needed in the main text. It represents a very condensed version of some chapters of the excellent book *Network flows* by R. K. Ahuja, T. L. Magnati and J. B. Orlin [10]. The content of the subsequent chapters is self-contained, so that it should be possible for the reader to understand the basic ideas of the various algorithms. In principle he should even be able to devise a particular implementation of one or the other code, although I recommend consulting existing public-domain (!) software libraries (e.g. [41]) first.

A The Maximum Flow/Minimum Cut Problem

A.1 Basic definitions

A **capacitated network** is a graph $G = (N, A)$, where N is the set of nodes and A the set of arcs, with *nonnegative* capacities u_{ij} (which can also be infinite) associated with each arc $(i, j) \in A$. In our first example of the random bond Ising model N is simply the set of lattice sites (plus two extra nodes, see Fig. 1), A the bonds between interacting sites and u_{ij} the ferromagnetic interaction strengths. Note that $u_{ij} \geq 0$ is essential. Let $n = \#N$ be the number of nodes in G and $m = \#A$ the number of arcs.

The **arc adjacency list** is the set of arcs emanating from a node: $A(i) = \{(i,j)|(i,j) \in A\}$.

One distinguishes two special nodes of N: the **source** node s and the **sink** node t.

A **flow** in the network G is a set of nonnegative numbers x_{ij} (or a map $\mathbf{x}: A \to R_+ \cup \{0, \infty\}$) subject to a **capacity constraint** for each arc

$$0 \leq x_{ij} \leq u_{ij} \quad \forall (i,j) \in A . \tag{35}$$

and to a **mass balance constraint** for each node

$$\sum_{\{j|(i,j)\in A\}} x_{ij} - \sum_{\{j|(j,i)\in A\}} x_{ji} = \begin{cases} -v & \text{for } i = s \\ +v & \text{for } i = t \\ 0 & \text{else} \end{cases} \quad (36)$$

This means that at each node everything that goes in has to go out, too, with the only exception being the source and the sink. What actually flows from s to t is v, the value of the flow.

The **maximum flow problem** for the capacitated network G is simply to find the flow **x** that has the maximum value v under the constraint (35) and (36).

We make a few assumptions: **1)** the network is directed, which means that for instance (i,j) is an arc pointing from node i to node j, **2)** whenever an arc (i,j) belongs to a network, the arc (j,i) also belongs to it or is added with zero capacity, **3)** all capacities are nonnegative integers, **4)** the network does not contain a directed path from node s to node t composed only of infinite capacity arcs, **5)** the network does not contain parallel arcs.[7]

A.2 Residual network and generic augmenting path algorithm

Now that we have defined the maximum flow problem, we have to introduce some tools with which it can be solved. The most important one is the notion of a *residual network*, which, as it is very often in mathematics, is already half the solution. If we have found a set of numbers x that fulfill the mass balance constraints, we would like to know whether this is already optimal, or on which arcs of the network we can improve (or *augment* in the jargon of combinatorial optimization) the flow.

Given a flow **x**, the **residual capacity** r_{ij} of any arc $(i,j) \in A$ is the maximum additional flow that can be sent from node i to node j using the arcs (i,j) and (j,i). The residual capacity has two components: 1) $u_{ij}-x_{ij}$, the unused capacity of arc (i,j), 2) x_{ji} the current flow on arc (j,i), which we can cancel to increase the flow from node i to j.

$$r_{ij} = u_{ij} - x_{ij} + x_{ji} \quad (37)$$

The **residual network** $G(\mathbf{x})$ with respect to the flow **x** consists of the arcs with *positive* residual capacities.

An **augmenting path** is a directed path from the node s to the node t in the residual network. The *capacity of an augmenting path* is the minimum residual capacity of any arc in this path.

Obviously, whenever there is an augmenting path in the residual network $G(\mathbf{x})$ the flow **x** is not optimal. This motivates the following generic augmenting path algorithm.

---

[7] All of these assumptions can be fulfilled in the physical problems we consider by appropriate modifications. E.g. number 3) can be fulfilled by rescaling the bond strengths J_{ij} with a factor and chopping off the decimal digits.

```
algorithm augmenting path
begin
    x:=0
    while G(x) contains a directed path from node s to t do
    begin
        identify an augmenting path P from node s to node t
        δ = min{r_ij|(i,j) ∈ P}
        augment δ units of flow along P and update G(x)
    end
end
```

Later we will see that the flow is indeed maximal if there is no augmenting path left. The main task in an actual implementation of this algorithm would be the identification of the directed paths from s to t in the residual network. Before we come to this point we have to make the connection to the minimum cut problem that is relevant for the physical problems discussed in the main text.

A.3 Cuts, labeling algorithm and max-flow-min-cut theorem

A **cut** is a partition of the node set N into two subsets S and $\overline{S} = N \backslash S$ denoted by $[S, \overline{S}]$. We refer to a cut as a *s-t-cut* if $s \in S$ and $t \in \overline{S}$.
The **forward** arcs of the cut $[S, \overline{S}]$ are those arcs $(i,j) \in A$ with $i \in S$ and $j \in \overline{S}$, the **backward** arcs those with $j \in S$ and $i \in \overline{S}$. The set of all forward arcs of $[S, \overline{S}]$ is denoted $(S, \overline{S})$.
The **capacity** of an *s-t*-cut is defined as $u[S, \overline{S}] = \sum_{(i,j) \in (S,\overline{S})} u_{ij}$. Note that the sum is only over forward arcs of the cut.
The **minimum cut** is a *s-t*-cut whose capacity u is minimal among all *s-t*-cuts.

Let $\mathbf{x}$ be a flow, v its value and $[S, \overline{S}]$ an *s-t*-cut. Then, by adding the mass balances for all nodes in S we have

$$v = \sum_{i \in S} \left\{ \sum_{\{j|(i,j) \in A(i)\}} x_{ij} - \sum_{\{j|(j,i) \in A(i)\}} x_{ji} \right\} = \sum_{(i,j) \in (S,\overline{S})} x_{ij} - \sum_{(i,j) \in (\overline{S},S)} x_{ji}. \quad (38)$$

Since $x_{ij} \leq u_{ij}$ and $x_{ji} \geq 0$ the following inequality holds

$$v \leq \sum_{(i,j) \in (S,\overline{S})} u_{ij} = u[S, \overline{S}] \quad (39)$$

Thus the value of any flow $\mathbf{x}$ is less or equal to the capacity of any cut in the network. If we discover a flow $\mathbf{x}$ whose value is equal to the capacity of some cut $[S, \overline{S}]$, then $\mathbf{x}$ is a maximum flow and the cut is a minimum cut. The following implementation of the augmenting path algorithm constructs a flow whose value is equal to the capacity of an *s-t*-cut it defines simultaneously. Thus it will solve the maximum flow problem (and, of course, the minimum cut problem).

As we have mentioned, our task is to find augmenting paths in the residual network. The following **labeling algorithm** uses a search technique to identify a directed path in $G(\mathbf{x})$ from the source to the sink. The algorithm fans out from the source node to find all nodes that are reachable from the source along a directed path in the residual network. At any step the algorithm has partitioned the nodes in the network into two groups: *labeled* and *unlabeled*. The former are those that the algorithm was able to reach by a directed path from the source, the latter are those that have not been reached yet. If the sink becomes labeled the algorithm sends flow along a path (identified by a predecessor list) from s to t. If all labeled nodes have been scanned and it was not possible to reach the sink, the algorithm terminates.

```
algorithm labeling
begin
   label node t
   while node t is labeled do
   begin
      unlabel all nodes
      set pred(j) = 0 for each j ∈ N
      label node s and set list := {s}
      while list ≠ ∅ and node t is unlabeled do
      begin
         remove a node i from list
         for each arc (i, j) ∈ A(i) in the residual network do
            if r_ij > 0 and node j is unlabeled then
               set pred(j) = i
               label node j
               add node j to list
      end
      if node t is labeled then augment
   end
end

procedure augment
begin
   Use the predecessor labels to trace back from the sink to
   the source to obtain an augmenting path P from s to t
   δ = min {r_ij|(i, j) ∈ P}
   augment δ units of flow along P, update residual capacities
end
```

Note that in each iteration the algorithm either performs an augmentation or terminates because it cannot label the sink. In the latter case the current flow is a maximum flow. To see this, suppose that at this stage S is the set of labeled nodes and $\overline{S} = N \backslash S$ is the set of unlabeled nodes. Clearly $s \in S$ and $t \in \overline{S}$. Since the algorithm cannot label any node in $\overline{S}$ from any node in S, $r_{ij} = 0$ for each $(i, j) \in (S, \overline{S})$, which implies with (37) $x_{ij} = u_{ij} + x_{ji}$. Thus $x_{ij} = u_{ij}$ (since

$0 \leq x_{ij} \leq u_{ij}$) for all $(i,j) \in (S, \overline{S})$ and $x_{ij} = 0$ for all $(i,j) \in (\overline{S}, S)$. Hence

$$v = \sum_{(i,j)\in(S,\overline{S})} x_{ij} - \sum_{(i,j)\in(\overline{S},S)} x_{ij} = \sum_{(i,j)\in(S,\overline{S})} x_{ij} = u[S,\overline{S}] \:. \tag{40}$$

This means that the flow **x** equals the capacity of the cut $[S, \overline{S}]$, and therefore **x** is a **maximum flow** and $[S, \overline{S}]$ is a **minimum cut**.

From these observation let us make the following conclusions:

Max-flow-min-cut theorem: The maximum value of the flow from a source node s to a sink node t in a capacitated network equals the minimum capacity among all s-t-cuts.

Augmenting path theorem: A flow $\mathbf{x}^*$ is a maximum flow if and only if the residual network $G(\mathbf{x}^*)$ contains no augmenting path.

Integrality theorem: If all arc capacities are integer, the maximum flow problem has an integer maximum flow.

Let n be the number of nodes, m the number of arcs and $U = \max\{u_{ij}\}$. Since any arc is examined at most once and the cut capacity is at most nU the complexity of this algorithm is $\mathcal{O}(nmU)$ (note that the flow increases at least by 1 in each augmentation). Because of the appearance of the number U it is a pseudo-polynomial algorithm. The so called preflow-push algorithms we discuss are now much more efficient, in particular they avoid the notorious delay caused by some bottleneck situations.

A.4 Preflow-push algorithm

The inherent drawback of the augmenting path algorithms is the computationally expensive operation of sending flow along a path, which requires $\mathcal{O}(n)$ time in the worst case. The preflow-push algorithms push flow on individual arcs instead of augmenting paths. They do not satisfy the mass balance constraints at intermediate stages. This is a very general concept in combinatorial optimization: algorithms either can operate within the space of admissible solutions and try to find optimality during iteration, or they can fulfill some sort of optimality criterion all the time and strive for feasibility. Augmenting path algorithms are of the first kind, preflow-push algorithms of the second. The basic idea is to flood the network from the source pushing as much flow as the arc capacities allow into the network towards the sink and then reduce it successively until the mass balance constraints are fulfilled.

A **preflow** is a function $\mathbf{x} : A \to R$ that satisfies the flow bound constraint $x_{ij} \leq u_{ij}$ and the following relaxation for the **excess** $e(i)$ of each node i:

$$e(i) := \sum_{\{j|(j,i)\in A\}} x_{ji} - \sum_{\{j|(i,j)\in A\}} x_{ij} \geq 0 \quad \forall i \in N\backslash\{s,t\} \:. \tag{41}$$

We have $e(t) \geq 0$ and only $e(s) < 0$. One denotes a node i to be **active** if its excess is strictly positive $e(i) > 0$.

As mentioned, preflow-push algorithms strive to achieve feasibility. Active nodes indicate that the solution is not feasible. Thus the basic operation of the algorithm is to select an active node and try to remove its excess by pushing flow to its neighbors. Since ultimately we want to send flow to the sink, we push flow to the nodes that are *closer* to the sink. This necessitates the use of distance labels:

- We say that a **distance function** $d : N \to Z_+ \cup \{0\}$ is **valid** with respect to a flow **x**, if it satisfies
  1. $d(t) = 0$ and
  2. $d(i) \leq d(j) + 1$ for every arc (i, j) in the residual network $G(\mathbf{x})$.
- If $d(\cdot)$ is valid then it also has the following properties (where n is the number of nodes):
  1. $d(i) \leq$ length of the shortest directed path from node i to t in $G(\mathbf{x})$
  2. $d(s) \geq n \Rightarrow G(\mathbf{x})$ contains no directed path from s to t.
- Furthermore we say that $d(\cdot)$ is **exact** if in 1) the equality holds.
- Finally an arc (i, j) is **admissible** if $d(i) = d(j) + 1$.

In the preflow-push algorithm we push flow on these admissible arcs. If the active node that we are currently considering has no admissible arcs, we increase its distance label so that we create at least one admissible arc.

```
algorithm preflow-push
begin
    preprocess
    while the network contains an active node do
    begin
        select an active node i
        push/relabel(i)
    end
end

procedure preprocess
begin
```
$\mathbf{x} := 0$
```
    compute the exact distance labels d(i)                    (1)
```
$x_{sj} = u_{sj}$ for each arc $(s, j) \in A$
$d(s) = n$
```
end

procedure push/relabel(i)
begin
    if the network contains an admissible arc (i,j) then
```
push $\delta = \min\{e(i), r_{ij}\}$ units of flow from node i to j
```
    else
```
replace $d(i)$ by $\min\{d(j) + 1 | (i, j) \in A \text{ and } r_{ij} > 0\}$
```
end
```

Ad (1): To compute the exact distance labels we have to calculate the shortest distances from node t to every other node, which we learn how to do in the next section.

The algorithm terminates when the excess resides at the source or at the sink, implying that the current preflow is a *flow*. Since $d(s) = n$ after peprocessing, and $d(i)$ is never decreased in push/relabel(i) for any i, the residual network contains no path from s to t, which means according to 2) above that there is no augmenting path. Thus the flow is maximal.

As in the context of the max-flow-min-cut theorem of the last section it might also be instructive here to visualize the generic preflow-push algorithm in terms of a network of (now flexible) water pipes representing the arcs with joints being the nodes. The distance function, which is so essential in this algorithm, measures how far nodes are above the ground. In this network we wish to send water from the source to the sink. In addition we visualize flow in admissible arcs as water flowing downhill. Initially, we move the source node upward, and water flows to its neighbors. In general, water flows downwards to the sink; however, occasionally flow becomes trapped locally at a node that has no downhill neighbors. At this point we move the node upward, and again water flows downhill to the sink. Eventually, no more flow can reach the sink. As we continue to move nodes upward, the remaining excess flow eventually flows back towards the source. The algorithm terminates when all the water flows either into the sink or flows back to the source.

The complexity of this algorithm turns out to be $\mathcal{O}(n^2 m)$, the so called FIFO preflow-push algorithm, which we do not discuss here, has a complexity of $\mathcal{O}(n^3)$.

B   Shortest Path Problems

B.1   Dijkstra's algorithm

Given a network $G = (N, A)$ and for each arc $(i, j) \in A$ a non-negative arc-length c_{ij}. In the above problem, where we had to find the exact distance labels in the preflow-push algorithm it is simply $c_{ij} = 1$ for all arcs in the residual network.

The task is to find the shortest paths from one particular node s to all other nodes in the network. *Dijkstra's algorithm* is a typical label-setting algorithm to solve this problem (with complexity $\mathcal{O}(n^2)$. It provides distance labels $d(i)$ with each node. Each of these is either temporarily (defining a set S) or permanently (defining a set $\overline{S} = N\backslash S$) labeled during the iteration, and as soon as a node is permanently labeled, $d(i)$ is the shortest distance. The path itself is reconstructed via predecessor indices.

First note that $d(j) = d(i) + c_{ij}$ for each arc (i,j) in a shortest path from node s to node i, and that $d(j) \geq d(i) + c_{ij}$ otherwise. By fanning out from node s the algorithm uses this criterion to find successively the shortest paths.

algorithm Dijkstra
begin
    $S := \emptyset$, $\overline{S} = N$
    $d(i) := \infty$ for each node $i \in N$
    $d(s) := 0$ and $pred(s) := 0$
    while $|S| < n$ **do**
    begin
        let $i \in \overline{S}$ be a node for which $d(i) = \min\{d(j)|j \in \overline{S}\}$
        $S := S \cup \{i\}$, $\overline{S} := \overline{S} \setminus \{i\}$
        for each $(i,j) \in A(i)$ do
            if $d(j) > d(i) + c_{ij}$ then
                $d(j) := d(i) + c_{ij}$ and $pred(j) := i$
    end
end

The fact that we always add a node $i \in \overline{S}$ with *minimal* distance label $d(i)$ ensures that $d(i)$ is indeed a shortest distance (there might be other shortest paths, but none with a strictly shorter distance). There are special implementations of this algorithm that have a much better running time than $\mathcal{O}(n^2)$.

B.2 Label correcting algorithm

As we said, Dijkstra's algorithm is a label-setting algorithm. The above mentioned criterion

$$d(i) \text{ shortest path distances } \Leftrightarrow d(j) \leq d(i) + c_{ij} \; \forall \, (i,j) \in A$$

gives also rise to a so called *label-correcting* algorithm.
Let us define reduced arc length (or **reduced costs**) via

$$c^d_{ij} := c_{ij} + d(i) - d(j) \; . \tag{42}$$

As long as one reduced arc length is negative, the distance labels $d(i)$ are not shortest path distances:

$$d(\cdot) \text{ shortest path distances } \Leftrightarrow c^d_{ij} \geq 0 \quad \forall (i,j) \in A \tag{43}$$

For later reference we also note the following observation. For any directed cycle W one has

$$\sum_{(i,j) \in W} c^d_{ij} = \sum_{(i,j) \in W} c_{ij} \tag{44}$$

The criterion (43) suggests the following algorithm for the shortest path problem:

```
algorithm label-correcting
begin
    d(s) := 0 and pred(s) := 0
    d(j) := ∞ for each node j ∈ N\{s}
    while some arc (i,j) satisfies d(j) > d(i) + c_ij  (c^d_ij < 0) do
    begin
        d(j) := d(i) + c_ij  (⇒ c^d_ij = 0)
        pred(j) = i
    end
end
```

The generic implementation of this algorithm has a running time $\mathcal{O}(\min\{n^2 mC, m2^n\})$ with $C = \max |c_{ij}|$, which is pseudo-polynomial. A FIFO implementation has complexity $\mathcal{O}(nm)$.

This algorithm also works for the cases in which some costs c_{ij} are negative, provided there are *no negative cycles*, i.e. closed directed paths W with $\sum_{(i,j)\in W} c_{ij} < 0$. In that case the instruction $d(j) := d(i) + c_{ij}$ would decrease some distance labels ad (negative) infinitum.

If there *are* negative cycles, one can detect them with an appropriate modification of the above code: One can terminate if $d(k) < -nC$ for some node k (again $C = \max |c_{ij}|$) and obtain these negative cycles by tracing them through the predecessor indices starting at node k. This will be useful in the next section.

C Minimum Cost Flow Problems

C.1 Definition

The next flow problem we discuss combines features of the maximum-flow and the shortest paths problem. The algorithm that solves it therefore also makes use of the ideas we presented so far. Let $C = (N, A)$ be a directed network with a *cost* c_{ij} and a *capacity* u_{ij} associated with every arc $(i, j) \in A$. Moreover we associate with each node $i \in N$ a number $b(i)$ which indicates its *supply* or *demand* depending on whether $b(i) > 0$ or $b(i) < 0$. The *minimum cost flow problem* is

$$\text{Minimize} \quad z(\mathbf{x}) = \sum_{(i,j)\in A} c_{ij} x_{ij} \tag{45}$$

subject to the mass balance constraints

$$\sum_{\{j|(i,j)\in A\}} x_{ij} - \sum_{\{j|(j,i)\in A\}} x_{ji} = b(i) \quad \forall i \in N \tag{46}$$

and the capacity constraints

$$0 \leq x_{ij} \leq u_{ij} \quad \forall (i,j) \in A \tag{47}$$

Again we make a few assumptions: **1)** All data (cost, supply/demand, capacity) are integral[8], **2)** the network is directed, **3)** $\sum_i b(i)$ and the minimum cost flow problem has a feasible solution (that means, one can find a flow x_{ij} that fulfills the mass balance and capacity constraints[9], **4)** there exists an uncapacitated directed path between every pair of nodes, **5)** all arc costs are non-negative (otherwise one could appropriately define a revered arc).

Again the **residual network** $G(\mathbf{x})$ corresponding to a flow $\mathbf{x}$ will play an essential role. It is defined in the same way as in the maximum flow problem, in addition the costs for the backwards arcs are reversed: a flow x_{ij} on arc $(i,j) \in A$ with capacity u_{ij} and cost c_{ij} will give rise to the arcs (i,j) and (j,i) with residual capacities $r_{ij} = u_{ij} - x_{ij}$ and $r_{ji} = x_{ij}$, respectively, and costs c_{ij} and $-c_{ij}$, respectively.

C.2 Negative cycle canceling algorithm

First we formulate a very important intuitive optimality criterion, the **negative cycle optimality criterion**: A feasible solution $\mathbf{x}^*$ is an *optimal* solution of the minimum cost flow problem, *if and only if* the residual network $G(\mathbf{x}^*)$ contains no negative cost cycle.

The proof is easy: Suppose the flow $\mathbf{x}$ is feasible and $G(\mathbf{x})$ contains a negative cycle. Then a flow augmentation along this cycle improves the function value $z(\mathbf{x})$, thus $\mathbf{x}$ is not optimal. Now suppose that $\mathbf{x}^*$ is feasible and $G(\mathbf{x}^*)$ contains no negative cycles and let $\mathbf{x}^0 \neq \mathbf{x}^*$ be an optimal solution. Now decompose $\mathbf{x}^0 - \mathbf{x}^*$ into augmenting cycles, the sum of the costs along these cycles is $\mathbf{c} \cdot \mathbf{x}^0 - \mathbf{c} \cdot \mathbf{x}^*$. Since $G(\mathbf{x}^*)$ contains no negative cycles $\mathbf{c} \cdot \mathbf{x}^0 - \mathbf{c} \cdot \mathbf{x}^* \geq 0$, and therfore $\mathbf{c} \cdot \mathbf{x}^0 = \mathbf{c} \cdot \mathbf{x}^*$ because optimality of $\mathbf{x}^*$ implies $\mathbf{c} \cdot \mathbf{x}^0 \leq \mathbf{c} \cdot \mathbf{x}^*$. Thus $\mathbf{x}^0$ is also optimal.

The following algorithm iteratively cancels negative cycles until the optimal solution is reached.

```
algorithm cycle canceling
begin
    establish a feasible flow x                                  (1)
    while G(x) contains a negative cycle do
    begin
        use some algorithm to identify a negative cycle W        (2)
        δ := min {r_ij|(i,j) ∈ W}
        augment δ units of flow in the cycle and update G(x)
    end
end
```

Ad **(1)**: Although, as we mentioned, in the usual physical problems a feasible solution is obvious in most cases (e.g. $\mathbf{x} = 0$) we note that, in principle, one has

---

[8] Here the same remark holds as for the maximum flow problem, previous footnote.
[9] In the physical models we discuss it is $b(i) = 0$ anyway, implying $\mathbf{x} = 0$ as a feasible solution.

to solve a maximum flow problem here: One introduces two extra-nodes s and t (source and sink, of course) and
$\forall i : b(i) > 0$ add a source arc (s, i) with capacity $u_{si} = b(i)$
$\forall i : b(i) < 0$ add a sink arc (i, t) with capacity $u_{it} = -b(i)$.
If the maximum flow from s to t saturates all source arcs (remember $\sum_i b(i) = 0$) the minimum cost flow problem is feasible and the maximum flow $\mathbf{x}$ is a feasible flow.

Ad **(2)**: For negative cycle detection in the residual network $G(x)$ one can use the label-correcting algorithm for the shortest path problem presented in the last section.

The running time of this algorithm is $\mathcal{O}(mCU)$, where $C = \max |c_{ij}|$ and $U = \max u_{ij}$, which means that it is pseudopolynomial. In the next section we present an alternative and more efficient way to solve the minimum cost flow problem.

C.3 Reduced cost optimality

Remember that when we considered the shortest path problem we introduced the reduced costs and obtained the shortest path optimality condition $c_{ij}^d = c_{ij} + d(i) - d(j) \geq 0$. This means

- c_{ij}^d is an optimal "reduced cost" for arc (i, j) in the sense that it measures the cost of this arc relative to the shortest path distances.
- With respect to the optimal distances, every arc has a nonnegative cost.
- Shortest paths use only zero reduced cost arcs.
- Once we know the shortest distances, the shortest path problem is easy to solve: Simply find a path from node s to every other node using only zero reduced cost arcs.

The natural question arises whether there is a similar set of conditions for more general min-cost flow problems. The answer is yes, as we show in the following. For the network defined in the last section associate a real number $\pi(i)$, unrestricted in sign with each node i, $\pi(i)$ is the **potential** of node i.
We define the **reduced cost** of arc (i, j) of a set of node potentials $\pi(i)$

$$c_{ij}^\pi := c_{ij} - \pi(i) + \pi(j) \ . \tag{48}$$

The reduced costs in the *residual network* are defined in the same way as the costs, but with c_{ij}^π instead of c_{ij}.
We have
1) For any directed path P from k to l: $\sum_{(i,j) \in P} c_{ij}^\pi = \sum_{(i,j) \in P} c_{ij} + d(k) - d(l)$.
2) For any directed cycle W: $\sum_{(i,j) \in W} c_{ij}^\pi = \sum_{(i,j) \in W} c_{ij}$
This means that negative cycles with respect to c_{ij} are also negative cycles with respect to c_{ij}^π.

Now we can formulate the reduced cost optimality condition:

A feasible solution $\mathbf{x}^*$ is an optimal solution of the min-cost flow problem
$$\Leftrightarrow$$
$\exists \pi$, a set of node potentials that satisfy the reduced cost optimality condition
$$c_{ij}^{\pi} \geq 0 \qquad \forall (i,j) \text{ arc in } G(\mathbf{x}^*).$$

For the implication "$\Leftarrow$" suppose that $c_{ij}^{\pi} \geq 0 \,\forall (i,j)$. One immediately realizes that $G(\mathbf{x}^*)$ contains no negative cycles since for each cycle W one has $\sum_{(i,j)\in W} c_{ij} = \sum_{(i,j)\in W} c_{ij}^{\pi} \geq 0$. For the other direction "$\Rightarrow$" suppose that $G(\mathbf{x}^*)$ contains no negative cycles. Denote with $d(\cdot)$ the shortest path distances from node 1 to all other nodes. Hence $d(j) \leq d(i) + c_{ij} \; \forall (i,j) \in G(\mathbf{x}^*)$. Now define $\pi = -d$ then $c_{ij}^{\pi} = c_{ij} + d(i) - d(j) \geq 0$. Note how closely connected the shortest path problem is to the min-cost flow problem.

There is an intuitive economic interpretation of the reduced cost optimality condition. Suppose we interpret c_{ij} as the cost of transporting one unit of a commodity from node i to node j through arc (i,j) and $\mu(i) = -\pi(i)$ as the cost of *obtaining* it at i. Then $c_{ij} + \mu(i)$ is the cost of commdity at node j, if we obtain it at node i and transport it to node j via arc (i,j). The inequality $c_{ij}^{\pi} \geq 0 \Leftrightarrow \mu(j) \leq c_{ij} + \mu(i)$ says that the cost of commodity at node j is no more than obtaining it at i and sending it via (i,j) — it could be smaller via other paths.

C.4 Successive shortest path algorithm

With the concept of reduced costs we can now introduce the successive shortest path algorithm for solving the min-cost flow problem. The cycle canceling algorithm maintains feasibility of the solution at every step and attempts to achieve optimality. In contrast, the successive shortest path algorithm maintains optimality of the solution ($c_{ij}^{\pi} \geq 0$) at every step and strives to attain feasibility (with respect to the mass balance constraints).
A **pseudoflow** $\mathbf{x}: A \to R^+$ satisfies the capacity and non-negaivity constraints, but not necessarily the mass balance constraints.
The **imbalance** of node i is defined as

$$e(i) := b(i) + \sum_{\{j|(ji)\in A\}} x_{ji} - \sum_{\{j|(ji)\in A\}} x_{ij}. \tag{49}$$

If $e(i) > 0$ then we call $e(i)$ the **excess** of node i, If $e(i) < 0$ then we call it the **deficit**. $E = \{i|e(i) > 0\}$ and $E = \{i|e(i) < 0\}$ are the sets of excess and deficit nodes, respectively. Note that because of $\sum_{i\in N} e(i) = \sum_{i\in N} b(i) = 0$ we have $\sum_{i\in E} e(i) = -\sum_{i\in D} e(i)$.

Let the pseudoflow $\mathbf{x}$ satisfy the reduced cost optimality condition with respect to the node potential π and $d(\cdot)$ the shortest path distances from some node s to all the other nodes in the residual network $G(\mathbf{x})$ with c_{ij}^{π} as arc lengths. Therefore we have:
Lemma 1:
a) For the potential $\pi' = \pi - d$ we have $c_{ij}^{\pi'} \geq 0$, too.

b) $c_{ij}^{\pi'} = 0$ for all arcs (i,j) on shortest paths.

To see a) note that $d(j) \leq d(i) + c_{ij}^{\pi}$, thus $c_{ij}^{\pi'} = c_{ij} + (\pi(i) - d(i)) - (\pi(j) - d(j)) = c_{ij}^{\pi} + d(j) - d(i) \geq 0$. For b) replace only the inequality by an equality.

The following lemma is the basis of the subsequent algorithm: Make the same assumption as in Lemma 1. Now send flow along a shortest path from some node s to some other node k to obtain a new pseudoflow $\mathbf{x}'$.

Lemma 2:

$\mathbf{x}'$ also satisfies the reduced cost optimality conditions!

For the proof take π and π' as in Lemma 1 and let P be the shortest path from node s to node k. Because of part b) of Lemma 1 we have $\forall (i,j) \in P: c_{ij}^{\pi'} = 0$. Therefore $c_{ji}^{\pi'} = -c_{ij}^{\pi'} = 0$. Thus a flow augmentation on $(i,j) \in P$ might add (j,i) to the residual network, but $c_{ji}^{\pi'} = 0$, which means that the reduced cost optimality condition $c_{ji}^{\pi'} \geq 0$ is still fulfilled.

The strategy for an algorithm is now clear. By starting with a feasible solution that fulfills the reduced cost optimality condition one can iteratively send flow along the shortest paths from excess nodes to deficit nodes to finally also fulfill the mass balance constraints.

algorithm successive shortest paths
begin
    $\mathbf{x} := 0$ $(G(\mathbf{x}) = G)$ and $\pi := 0$ $(c_{ij} = c_{ij}^{\pi} \geq 0)$
    $e(i) := b(i)$ $\forall i \in N$
    $E := \{i | e(i) > 0\}$, $D := \{i | e(i) < 0\}$.
    while $E \neq \emptyset$ do
    begin
        select a node $k \in E$ and a node $l \in D$
        determine shortest path distance $d(j)$ from some node s to
            all other nodes in $G(\mathbf{x})$ w.r. to the reduced costs c_{ij}^{π}
        let P denote a shortest path from node k to node l
        update $\pi := \pi - d$
        $\delta := \min\{e(k), -e(l), \min\{r_{ij} | (i,j) \in P\}\}$
        augment δ units of flow along path P
        update $\mathbf{x}$, $G(\mathbf{x})$, E, D and the reduced costs
    end
end

Note that in each iteration one excess is decreased by increasing flow by at least one unit. Denoting with U the upper bound on the largest supply of any node one needs at most nU iterations, in each of which one has to solve a shortest path problem with non-negative arc lengths (so Dijkstra's algorithm is appropriate). This means that the above algorithm is polynomial if we know how U scales with m or n.

C.5 Convex cost flows

The cycle annealing algorithm as well as the successive shortest path algorithm solve the minimum cost flow problem for a *linear* cost function $\sum_{(i,j) \in E} c_{ij} \cdot x_{ij}$, where c_{ij} represents the cost for sending one unit of flow along the arc (i, j). This problem can be generalized to the following situation:

$$\text{Minimize} \quad z(\mathbf{x}) = \sum_{(i,j) \in \Lambda} h_{ij}(x_{ij}) \tag{50}$$

subject to the mass balance constraints (46) and the capacity constraint (47). In addition we demand the flow variables x_{ij} to be integer. The functions $h_{ij}(x_{ij})$ can be any non-linear function, which, however, has to be *convex*, i.e.

$$\forall x, y, \text{ and } \theta \in [0, 1] \quad h_{ij}(\theta x + (1 - \theta)y) \leq \theta h_{ij}(x) + (1 - \theta)h_{ij}(y) \tag{51}$$

For this reason it is called the *convex cost flow problem*. Without loss of generality we can assume that $h_{ij}(0) = 0$. Here the cost (for *one* unit) depends on the actual flow (since $h_{ij}(x_{ij})$ is a nonlinear function of the flow variable x_{ij}):

$$\begin{aligned} c_{ij}(x_{ij}) &:= h_{ij}(x_{ij} + 1) - h_{ij}(x_{ij}) \\ c_{ji}(x_{ij}) &:= h_{ij}(x_{ij} - 1) - h_{ij}(x_{ij}) \end{aligned} \tag{52}$$

Now c_{ij} and c_{ji} are the costs for increasing and decreasing, respectively, the the flow variable x_{ij} by one.

After introducing these quantities it becomes straightforward to solve this problem with slight modifications of the algorithms we already have at hand. The first is again a negative cycle canceling algorithm:

```
algorithm cycle canceling (convex costs)
begin
    establish a feasible flow x
    calculate the costs c(x) as in (52)
    while [G(x), c(x)] contains a negative cycle do
    begin
        use some algorithm to identify a negative cycle W
        augment one unit of flow in the cycle
        update G(x) and c(x)
    end
end
```

Note that since $h_{ij}(x)$ is convex the cost for augmenting x by *more* than one unit *increases* the costs. This ensures that if we do not find any negative cycles, the flow is indeed optimal.

This algorithm is, unfortunately non-polynomial in time, although it performs reasonably well on average. The successive shortest path algorithm discussed in the last section can also be applied in the present context with a significant gain in efficiency. For this algorithm it was essential that the reduced costs c_{ij}^π with respect to some node potential π maintains the reduced cost optimality condition $c_{ij}^\pi \geq 0$ upon flow augmentation along shortest paths. Now the question is, whether this still holds if with the change of the flow (caused by the augmentation) the costs also change. To show this we prove the following

Lemma: Let s be an excess node, $d(\cdot)$ shortest path distances w.r.t. the reduced costs c_{ij}^π from node s to all other nodes, $\pi' = \pi + d$, t a deficit node, W a shortest path from s to t, and $\mathbf{x}^{\text{new}}$ the flow that one obtains by augmenting $\mathbf{x}$ along W by *one* unit. Then:

$$c_{ij}^{\pi'} \geq 0 \text{ also for the } \textit{modified arc costs} \text{ along } W.$$

For the proof let $w_{ij} \in \{+1, -1\}$ for an arc $(i,j) \in W$ with $x_{ij}^{\text{new}} = x_{ij} + w_{ij}$. Then the modified costs on this arc are

$$c_{ij}^* = h_{ij}(x_{ij} + 2w_{ij}) - h_{ij}(x_{ij} + w_{ij}) \geq h_{ij}(x_{ij} + w_{ij}) - h_{ij}(x_{ij}) = c_{ij}\ .$$

because of the convexity of $h_{ij}(x)$. From this $c_{ij}^{\pi'*} = c_{ij}^{\pi'} + \pi'(i) - \pi'(j) \geq c_{ij} + \pi'(i) - \pi'(j)$ follows for the modified reduced costs, which proves the lemma.

Thus we have the successive shortest path algorithm for the convex costs flow problem:

algorithm successive shortest paths (convex costs)
```
begin
   x := min {H(x) | x ∈ Z^A} and π := 0
   e(i) := b(i) + ∑_{j|(ji)∈A} x_ji - ∑_{j|(ji)∈A} x_ij  ∀i ∈ N
   while there is a node s with e(s) > 0 do
   begin
      compute the reduced costs c^π(x)
      determine shortest path distance d(·) from s to
         all other nodes in G(x) w.r. to the reduced costs c_ij^π
      choose a node t with e(t) < 0
      augment x along shortest path from s to t by one unit
      π = π - d
   end
end
```

Note that we start with an optimal solution for $H(\mathbf{x})$, i.e. for each arc (i,j) we choose the value of x_{ij} in such a way that it is minimal. By this we guarantee that $c_{ij}(x_{ij}) \geq 0$ and thus that the reduced costs c_{ij}^π with $\pi = 0$ fulfill the optimality condition $c_{ij}^\pi \geq 0$. The complexity of this algorithm is strictly polynomial in the physical example we discuss in section 8 [31].

Final Remark

For everybody who encounters one of the network flow problems mentioned above in his study of a physical (or any other) problem an important piece of advice: Before reinventing the wheel, that is before wasting time in hacking a subroutine that solves a particular network flow problem, one should consult the extremely valuable LEDA (library of efficient data types and algorithms) library, where many source codes of highly efficient combinatorial optimization algorithms can be found. All information, the manual [41] and the source codes

can be found on the internet (this is public domain):
    http://www.mpi--sb.mpg.de/LEDA/leda.html
Have fun!

References

[1] Toulouse, G., *Commun. Phys.* **2** 115 (1977).
[2] Villain, J., *J. Phys.* **C 10** 1717 (1977); Forgacs, G., *Phys. Rev.* **B 22** 4473 (1980); Harris, A. B., Kallin, C., Berlinsky, A. J., *Phys. Rev.* **B 45** 2899 (1992); Huse D. A., Rutenberg, A. D., *Phys. Rev.* **B 45** 7536 (1992); Chalker, J. T., Holdsworth, P. C. W., Shender, E. F., *Phys. Rev. Lett.* **68** 855 (1992); Shore, J. D., Holzer, M., Sethna, J. P., *Phys. Rev.* **B 46** 11376 (1992); Chandra, P., Coleman, P., Ritchey, I., *J. Physique* **I 3** 591 (1993); Reimers, J. N., Berlinsky, A. J., *Phys. Rev.* **B 48** 9539 (1993).
[3] *The traveling salesman problem*, ed.: Lawler, E. L., Lenstra, J. K., Rinnooy Kan, A. H. G., Shmoys, D.,B., Wiley-Interscience series in discrete mathematics, (John Wiley & Sons, Chichester, 1985).
[4] Wilson, R. J., *Introduction to graph theory*, (Oliver and Boyd, Edinburgh, 1972); Bondy, J. A., Murty, U. S. R., *Graph theory with applications*, (Mac Millan, London, 1976).
[5] Lawler, E. L., *Combinatorial optimization: Networks and matroids*, (Holt, Rinehart and Winston, New York, 1976).
[6] Papadimitriou C. H., Steiglitz, K., *Combinatorial optimization: Algorithms and complexity*, (Prentice-Hall, Englewood Cliffs NJ, 1982).
[7] Chvátal, V., *Linear programming*, (Freeman, San Francisco, 1983).
[8] Derigs, U., *Programming in networks and graphs*, Lecture Notes in Economics and Mathematical Systems **300**, (Springer-Verlag, Berlin-Heidelberg, 1988).
[9] Grötschel, M., Lovász, L., Schrijver, A., *Geometric algorithms and combinatorial optimization*, (Springer-Verlag, Berlin-Heidelberg, 1988).
[10] Ahuja, R. K., Magnati, T. L., Orlin, J. B., *Network Flows*, (Prentice Hall, London, 1993).
[11] Middleton, A. A., *Phys. Rev.* **E 52** R3337 (1995).
[12] Rieger, H., *Monte Carlo simulations of Ising spin glasses and random field systems* in *Annual Reviews of Computational Physics II*, p. 295–341, (World Scientific, Singapore, 1995).
[13] Barahona, F., *J. Phys.* **A 18** L673 (1985).
[14] Ogielski, A. T. *Phys. Rev. Lett.* **57** 1251 (1986).
[15] Fishman. S., Aharony, A., *J. Phys.* **C 12** L729 (1979).
[16] Hartmann A. K., Usadel, K. D., *Physica* **A 214** 141 (1995).
[17] Esser, J., private communication.
[18] Efros. A. L., Shklovskii, B. I., *J. Phys.* **A 8** L49 (1975).
[19] Tenelsen, K., Schreiber, M., Phys. Rev. **B 49**, 12622 (1994); **52**, 13287 (1995).

[20] Barahona, F. *J. Phys.* **A 15**, 3241 (1982); Barahona, F., Maynard, R., Rammal, R., Uhry, J. P., *J. Phys.* **A 15**, 673 (1982).
[21] Kawashima N., Rieger, H., submitted to Europhys. Lett., cond-mat/9612116.
[22] Grötschel, M., Jünger, M., Reinelt, G., in: *Heidelberg Colloqium on Glassy dynamics and Optimization* , ed. L. van Hemmen and I. Morgenstern (Springer-Verlag, Heidelberg 1985).
[23] De Simone, C., Diehl, M., Jünger, M., Mutzel, P., Reinelt, G., Rinaldi, G., *J. Stat. Phys.* **80** 487 (1995).
[24] Klotz, T., Kobe, S., *J. Phys.* **A 27** L95 (1994).
[25] Diehl, M., *Determination of exact ground states of Ising spin glasses with a branch-and-cut algorithm*, (Diploma Thesis, Köln, 1995) unpublished, postscript file available at:
http://www.informatik.uni-koeln.de/ls_juenger/staff/diehl.html.
[26] Thienel, S., *ABACUS — A Branch–And–CUt System*, (Ph.D. thesis, Köln, 1995) unpublished, postscript file available at http://www.informatik.uni-koeln.de/ls_juenger/publications/thienel/diss.html
[27] Bray, A. J., Moore, M. A., in: *Heidelberg Colloqium on Glassy dynamics and Optimization*, ed. L. van Hemmen and I. Morgenstern (Springer-Verlag, Berlin-Heidelberg 1985).
[28] Rieger, H., Santen, L., Blasum, U., Diehl, M., Jünger, M., Rinaldi, G., *J. Phys.* **A 29** 3939 (1996).
[29] Maynard R., Rammal, R., *J. Phys. Lett.* (France) **43** L347(1982); Ozeki, Y., *J. Phys. Soc. Jpn.* **59** 3531 (1990).
[30] Shirakura, T., Matsubara, F., *J. Phys. Soc. Jpn.* **64** 2338 (1995); Ozeki, Y., Nonomura, Y., *J. Phys. Soc. Jpn.* **64** 3128 (1995).
[31] Blasum, U., Hochstättler, W., Moll, C., Rieger, H., *J. Phys.* **A 29** L459 (1996).
[32] Rieger, H., Blasum, U., *Phys. Rev. Lett.*, in press.
[33] Tsai, Y. C., Shapir, Y., *Phys. Rev. Lett.* **69** 1773 (1992); *Phys. Rev.* **E 40** (194) 3546, 4445.
[34] See e. g. Chui, T. S., Weeks, J., D., *Phys. Rev.* **B 14** 4978 (1976).
[35] Zeng, C., Middleton, A. A., Shapir, Y., *Phys. Rev. Lett.* **77** 3204 (1996).
[36] Kleinert, H., *Gauge Fields in Condensed Matters*, (World Scientific, Singapore, 1989).
[37] Rieger, H., Kisker, J., in preparation.
[38] Wengel, C., Young, A. P., *Phys. Rev.* **B 54** R6869 (1996).
[39] Bokil, H. S., and Young, A. P., *Phys. Rev. Lett.* **74** 3021 (1995).
[40] *Traffic and Granular Flow*, ed. D. E. Wolf, M. Schreckenberg and A. Bachem (World Scientific, Singapore 1996).
[41] Näher, S., Uhrig, C., *The LEDA User Manual, Version R3.4* (Martin–Luther Universität Halle–Wittenberg, Germany, 1996), postscript file available at http://www.mpi-sb.mpg.de/LEDA/leda.html.

Molecular Dynamics

Hans Herrmann

ICA1, Universität Stuttgart, D-70569 Germany

Abstract. Molecular dynamics is a deterministic technique to solve the equation of motion of a system of N particles:

$$m_i \dot{\mathbf{v}}_i = \mathbf{f}_i = m_i \mathbf{g} + \sum_j \mathbf{f}_{ij} \tag{1}$$

where m_i, $\mathbf{r}_i$ and $\mathbf{v}_i$ are mass, position and velocity of particle i with $i = 1, ..., N$ and where the force between particle i and j is given through $\mathbf{f}_{ij} = -\nabla v_{ij}$ by the pair potential v_{ij} which usually only depends on the distance r_{ij} between the centers of mass of the two particles.

1 Methods of Finite Differences

If the potential is not too steep ($\partial v_{ij}/\partial r_{ij}$ small) one can solve the equations by introducing a small fixed time step Δt. At each time step t one then calculates the forces $\mathbf{f}_i$ and obtains the accelerations through $\dot{\mathbf{v}}_\mathbf{i} = \mathbf{f}_i/m_i$ from which one obtains positions and velocities at time step $t+\Delta t$ using various possible schemes.

The quality of the scheme one uses is determined by the following criteria:
- The individual iteration step should be fast.
- Δt should be as large as possible.
- The deviations from the exact trajectories should be as small as possible.
- The memory requirements should not be too large.
- The programme should be simple.
- The valid conservation laws should be fulfilled (energy conservation, momentum conservation in the case of periodic boundary condition, etc).

2 Verlet and Leap-Frog Algorithms

If one adds the two Taylor expansions

$$\mathbf{r}_i(t \pm \Delta t) = \mathbf{r}_i(t) \pm \Delta t \mathbf{v}_i(t) + \frac{1}{2}\Delta^2 t \dot{\mathbf{v}}_\mathbf{i}(t) \tag{2}$$

one obtains:

$$\mathbf{r}_i(t + \Delta t) = 2\mathbf{r}_i(t) - \mathbf{r}_i(t - \Delta t) + \Delta^2 t \dot{\mathbf{v}}_\mathbf{i}(t) \tag{3}$$

which is the scheme introduced in 1967 by Verlet. The velocities can then be obtained through

$$\mathbf{v}_i(t) = \frac{\mathbf{r}_i(t + \Delta t) - \mathbf{r}_i(t - \Delta t)}{2\Delta t}. \tag{4}$$

The drawback of the above method is that one compares a term of order $\Delta^0 t$ with a term of order $\Delta^2 t$ which for small Δt is numerically not very convenient. This can be circumvented by the leap frog scheme in which one considers velocities at intermediate time steps:

$$\mathbf{v}_i(t + \frac{1}{2}\Delta t) = \mathbf{v}_i(t - \frac{1}{2}\Delta t) + \Delta t \dot{\mathbf{v}}_i(t) \tag{5}$$

and updates the positions through

$$\mathbf{r}_i(t + \Delta t) = \mathbf{r}_i(t) + \Delta t \mathbf{v}_i(t + \frac{1}{2}\Delta t). \tag{6}$$

3 The Predictor-Corrector Method

One can extrapolate the variables describing a particle at time $t + \Delta t$ through a Taylor expansion before calculating the forces and do this even more precisely by taking into account higher order time derivatives. A scheme of fourth order like

$$\mathbf{r}_i^p(t + \Delta t) = \mathbf{r}_i(t) + \Delta t \mathbf{v}_i(t) + \frac{1}{2}\Delta^2 t \dot{\mathbf{v}}_i(t) + \frac{1}{6}\Delta^3 t \ddot{\mathbf{v}}_i(t) + \ldots$$

$$\mathbf{v}_i^p(t + \Delta t) = \mathbf{v}_i(t) + \Delta t \dot{\mathbf{v}}_i(t) + \frac{1}{2}\Delta^2 t \ddot{\mathbf{v}}_i(t) + \ldots \tag{7}$$

$$\dot{\mathbf{v}}_i^p(t + \Delta t) = \dot{\mathbf{v}}_i(t) + \Delta t \ddot{\mathbf{v}}_i(t) + \ldots$$

$$\ddot{\mathbf{v}}_i^p(t + \Delta t) = \ddot{\mathbf{v}}_i(t) + \ldots$$

thus gives predicted values at $t + \Delta t$ and implies keeping the four variables $\mathbf{r}_i, \mathbf{v}_i, \dot{\mathbf{v}}_i$ and $\ddot{\mathbf{v}}_i$ for each particle. After calculating the force $\mathbf{f}_i$ with these predicted values one obtains the correct acceleration $\dot{\mathbf{v}}_i^c = \mathbf{f}_i/m_i$ and defining $\boldsymbol{\delta} = \dot{\mathbf{v}}_i^c - \dot{\mathbf{v}}_i^p$ one can obtain the correct positions, velocities, etc through

$$\mathbf{r}_i^c = \mathbf{r}_i^p + c_1 \boldsymbol{\delta}$$

$$\mathbf{v}_i^c = \mathbf{v}_i^p + c_2 \boldsymbol{\delta} \tag{8}$$

$$\ddot{\mathbf{v}}_i^c = \ddot{\mathbf{v}}_i^p + c_4 \boldsymbol{\delta}$$

where the c_j are the Gear coefficients which can been obtained through a minimalization principle. Their value depends on the order of the scheme and the order of the differential equation that one solves (2nd order in our case) and can be looked up in tables (e.g. in Appendix E of [1]). For (8) they are: $c_1 = 1/6, c_2 = 5/6$ and $c_4 = 1/3$.

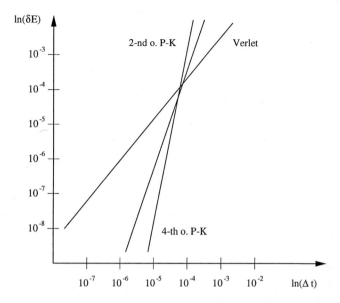

Fig. 1. Comparison of the accuracy of the Verlet, the 2-nd order and the 4-th order predictor-corrector methods

4 Comparison of the Schemes

A good criterion for the precision of a method is to calculate the energy fluctuations $\delta E = \sqrt{<E^2> - <E>^2}$. Let us consider a simulation of the system during a fixed time interval T for different Δt. That means we have to choose the number of iterations n such that $n = T/\Delta t$. In the figure above we see the precision δE one obtains empirically for different methods in a double logarithmic plot.

Since one wants to work at the largest possible Δt one can conclude from this figure that it is more convenient to use the Verlet algorithm if one is not interested in high precision but to use the predictor-corrector scheme if one needs high precision.

5 Tricks to Improve the Speed of the Algorithms

The most expensive part of the schemes discussed above is the calculation of the forces; on the one hand because most potentials are not simple functions and on the other because one has N^2 pairs of particle interactions.

In the calculation of the force itself one should avoid taking the square root when calculating the distance r_{ij} between i and j. This is easy when the potential is an even function of the distance since then only r_{ij}^2 is needed. Another good trick to avoid heavy number crunching in the case of complicated potentials is to tabulate them at the beginning of the programme into a vector and then just calculate the address in the vector and make a memory fetch inside the loop.

The more serious problem is, however, the N^2-loop. If the potential is not long ranged it can be set to zero beyond some cut-off distance r_c. Then two famous tricks are available: the Verlet tables and "linked cells". In the case of Verlet tables one introduces a further surrounding of radius $r_l > r_c$ around every particle. The indices of all the particles within these surroundings are stored in a long vector "LIST" such that first come the particles in the vecinity of particle 1 then those around particle 2 and so on. To find the particles in the vecinity of particle i one has a vector "POINT" which indicates which is the first particle in "LIST". The last particle in the vicinity of i therefore has the address POINT(i+1)-1 in LIST. The Verlet table has to be updated every n steps where n is given by the maximum velocity v_{max} of a particle in the system through $n = (r_l - r_c)/(v_{max} \Delta t)$. Therefore the algorithm still goes as N^2 (but with a smaller prefactor). Typical values for the Lenard-Jones potential: $v_{ij} = 4\epsilon((\sigma/r_{ij})^{12} - (\sigma/r_{ij})^6)$ are $r_c = 2.4\sigma, r_l = 2.7\sigma$ and $n = 20$.

To discuss the linked cell technique let us consider two dimensions for simplicity. One puts over the system a square grid of size $M \times M$ such that the length of a cell is larger than r_c. For a particle i in cell j interactions are only possible with particles within 9 cells: the cell j, the four nearest neighboring cells and the four next-neighboring cells. In order to find the indices of the particles within these cells one uses two vectors: "LIST(N)" and "BEG(M^2)". For each particle LIST contains the next particle in the same cell and is zero if there are no further particles in the cell. BEG(j) contains the first particle in cell j in that LIST. Example:

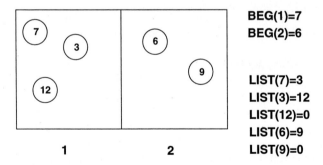

Fig. 2. Schematic illustration of the linked cell technique

The update of the linked cells can be done locally since it is easy to determine from the position $\mathbf{r}_i$ which cell a particle belongs in. Therefore one has reduced the number of possible pair interactions by a factor of $9/M^2$ and obtains a loop of order N instead of N^2. The linked cell structure is also used when one wants to parallelize the programme ([2]).

6   Canonical Simulations

Until now we have described microcanonical simulations. In most cases, however, one is interested in working at a fixed temperature T instead of a fixed energy. Various methods have been proposed to this end which will be presented briefly in their historical order.

A combination of molecular dynamics and Monte Carlo was proposed in 1980 by Andersen: Every n time steps Δt one picks at random a particle and gives to it a velocity v chosen from the Maxwell-Boltzmann distribution:

$$P(v) = \frac{\rho}{(2\pi mkT)^{\frac{3}{2}}} \exp -\frac{mv^2}{2kT}. \qquad (9)$$

For large n molecular dynamics dominates and for small n the configuration is dominated by the stochastic process. The disadvantage of this method is the loss of determinism and the arbitrariness of the parameter n which must be chosen empirically.

Another technique was proposed in 1982 and 1983, respectively, by Hoover and Evans which consists in keeping the temperature exactly constant. This can be achieved by rescaling all the velocities by a factor $\delta = \sqrt{T/\mathcal{T}}$ at each time step where the actual temperature $\mathcal{T}$ is defined through

$$k\mathcal{T} = \frac{1}{2} \sum_i m_i \mathbf{v}_i^2. \qquad (10)$$

Such a rescaling, however, can not be applied in a predictor-corrector scheme since the higher time derivatives must also be taken into account. A more general way to enforce a constant temperature uses a Lagrange multiplier ξ by considering the equation for the momenta

$$\dot{\mathbf{p}}_i = \mathbf{f}_i - \xi \mathbf{p}_i \qquad (11)$$

where through a minimalization principle one obtains

$$\xi = \frac{\sum_i \mathbf{p}_i \mathbf{f}_i}{\sum_i \mathbf{p}_i^2}. \qquad (12)$$

The temperature of the system is now the same as that of the initial configuration which is not always simple to impose. But the worst disadvantage of this method is that (11) does not describe the original equation of motion anymore.

In 1984 Nosé proposed coupling a heat bath through an additional scalar degree of freedom s which has a conjugate momentum p_s and a mass Q. Its kinetic and potential energy are given by $p_s^2/2Q$ and $dNkT\ln s$ respectively and the coupling to the particle variables is achieved through a rescaling of the time scale through: $\mathbf{v}_i = \mathbf{p}_i/sm_i$. By eliminating this degree of freedom from the resulting equations of motion again, Hoover formulated in 1985 the following system of equations

$$\dot{\mathbf{p}}_i = \mathbf{f}_i - \eta \mathbf{p}_i \qquad (13)$$

$$\dot{\eta} = \frac{1}{Q}\left(\sum_i \frac{\mathbf{p}_i^2}{2m_i} - dNkT\right)$$

also called the Nosé-Hoover thermostat. Q now plays the rôle of a relaxation parameter: For large Q the coupling to the heat bath is weak and for too small Q the coupling is so strong that it can lead to spurious oscillations. Again Q must be chosen empirically. Nevertheless today this is considered to be the best way to simulate at fixed temperature.

One can also work in other ensembles, for instance at constant pressure instead of constant volume by using a similar approach to that of (13) which physically corresponds to considering a system where the volume can be changed by a piston of mass Q.

7 Event Driven Molecular Dynamics

If the potential is very steep at some points, as is typically the case for hard core interactions, the above methods are not very efficient since they would require a very small Δt. In such a case it is more convenient to use a completely different approach: Instead of calculating forces, the collisions between particles are considered to be point-like and the velocities of the particles resulting after a binary collision are obtained from the initial velocities from a collision rule which is based on the physical effects of the collision such as energy or momentum conservation, surface slip or friction. Let us consider for simplicity that the particles i are hard core spheres of radius R_i and that no gravity acts. Since the trajectories are ballistic, as long as there is no collision one only needs to know the time t_c at which the next collision in the system occurs.

To calculate the collision time t_c let us first consider every pair of particles i and j. Their collision time t_{ij} is given by

$$\mid \mathbf{r}_{ij}(t+t_{ij}) \mid = \mid \mathbf{r}_{ij}(t) + \mathbf{v}_{ij}t_{ij} \mid = R_i + R_j \tag{14}$$

where $\mathbf{v}_{ij}$ is the relative velocity between the particles. Squaring (14) one gets

$$\mathbf{v}_{ij}^2 t_{ij}^2 + 2(\mathbf{v}_{ij}\mathbf{r}_{ij})t_{ij} + \mathbf{r}_{ij}^2 - (R_i + R_j)^2 = 0 \tag{15}$$

and from this quadratic equation one obtains t_{ij} exactly. The collision time is then given by

$$t_c = \min_{ij} t_{ij}. \tag{16}$$

This global minimization is of course fatal to any vectorization or parallelization attempt which is one of the major drawbacks of this method. Some ways to optimize the algorithm to prevent an N^2 operation were proposed by [3]. After t_c has been determined through (16) all particles are moved ballistically: $\mathbf{r}'_i = \mathbf{r}_i + \mathbf{v}_i t_c$, to the point at which the next collision occurs in the system.

The simplest conditions to calculate the velocities are momentum and energy conservation and the hypothesis of infinite slip, i.e. that all momentum transfer must be perpendicular to the tangential contact plane. Due to momentum conservation, the momentum transfer $\Delta \mathbf{p}$ to particle i is just the same as that to particle j except for a change of sign. The direction of $\Delta \mathbf{p}$ is given by the condition of infinite slip, i.e. is parallel to the vector $\mathbf{r}_{ij}$ connecting the centers of mass of the spheres. Energy conservation finally gives:

$$\Delta \mathbf{p} = -2m_{eff}(\mathbf{v}_{ij}\mathbf{r}_{ij})\frac{\mathbf{r}_{ij}}{r_{ij}^2} \qquad (17)$$

where the effective mass is defined as

$$m_{eff} = \frac{m_i m_j}{m_i + m_j}. \qquad (18)$$

In the case of tangential momentum transfer, in particular in the extreme case of zero slip, it is important to include the rotation of the particles, i.e. to add as variables the angle ϕ_i, the angular velocity ω_i and eventually higher order time derivatives. The local velocity at the contact point is then

$$\mathbf{v}_{rel} = \mathbf{v}_i - \mathbf{v}_j + (R_i \omega_i + R_j \omega_j)\mathbf{s} \qquad (19)$$

where $\mathbf{s}$ is the unit vector in the contact plane. For three dimensional spheres one then finds using the no slip condition and the conservation of angular momentum that the tangential momentum transfer is $\frac{2}{7}$ times the normal momentum transfer.

For macroscopic particles one also often has dissipation due to acoustic emission, plastic deformation or viscoelastic damping. This dissipation can be described by a material constant r called the restitution coefficient defined as $r = -v^f/v^i$ where v^f and v^i are the final and initial velocities of the particles after collision with a wall of the same material. In that case (17) is modified such that the 2 is replaced by $1 + r$ ([4]).

The efficiency of event driven simulations depends crucially on the average value of t_c. If t_c is much larger than the Δt of finite difference methods the disadvantages related to the global minimum search are compensated and the method is efficient. The most important parameter is the density of the system: for high densities t_c becomes too small so that at the end the computer spends all its time calculating tiny oscillations between close particles. In the case of dissipative particles McNamara and Young have found that one could spend infinitely many iterations to describe a collision of several particles occuring at a finite time (finite time singularity) ([5]).

8 Conclusion

We have seen that plenty of methods and computer tricks are available to simulate the Newtonian trajectories of an ensemble of atoms or grains. (An excellent book which contains nearly everything said here and many more useful details is [1].) While Peter Lomdahl has published simulations of 137 million particles and even larger systems have been simulated on the Paragon at Oak Ridge the problem lies more in making sufficiently many iterations to describe experimental times. In the case of finite difference methods Δt is given by the radius of the particles divided by the maximal velocity. For atoms this gives typically picoseconds so that even after thousands of iterations one is still in the range of nanoseconds. For macroscopoic grains like powders or sand Δt is of the order of $10^{-4} - 10^{-6}$ seconds so that it becomes possible to simulate minutes ([6]).

References

[1] Allen, M. P., Tildesley, D. J., *Computer Simulation of Liquids*. (Clarendon Press, Oxford, 1987).
[2] Ristow, G., *Int. J. Mod. Phys.* **C 3** 1281 (1992).
[3] Lubachevsky, B., *Int. J. in Computer Simulation* **2** 373 (1992).
[4] Luding, S., *Phys. Rev.* **E 52** 4442 (1995).
[5] McNamara, S., Young, W. R., *Phys. Fluid.* **A 5** 34 (1993)
[6] Herrmann, H. J., in *3rd Granada Lectures on Computational Physics*, eds. P.L. Garrido and J. Marro (Springer, Heidelberg, 1995, p. 67-114).

Lecture Notes in Physics

For information about Vols. 1–465
please contact your bookseller or Springer-Verlag

Vol. 466: H. Ebert, G. Schütz (Eds.), Spin – Orbit-Influenced Spectroscopies of Magnetic Solids. Proceedings, 1995. VII, 287 pages, 1996.

Vol. 467: A. Steinchen (Ed.), Dynamics of Multiphase Flows Across Interfaces. 1994/1995. XII, 267 pages. 1996.

Vol. 468: C. Chiuderi, G. Einaudi (Eds.), Plasma Astrophysics. 1994. VII, 326 pages. 1996.

Vol. 469: H. Grosse, L. Pittner (Eds.), Low-Dimensional Models in Statistical Physics and Quantum Field Theory. Proceedings, 1995. XVII, 339 pages. 1996.

Vol. 470: E. Martí'nez-González, J. L. Sanz (Eds.), The Universe at High-z, Large-Scale Structure and the Cosmic Microwave Background. Proceedings, 1995. VIII, 254 pages. 1996.

Vol. 471: W. Kundt (Ed.), Jets from Stars and Galactic Nuclei. Proceedings, 1995. X, 290 pages. 1996.

Vol. 472: J. Greiner (Ed.), Supersoft X-Ray Sources. Proceedings, 1996. XIII, 350 pages. 1996.

Vol. 473: P. Weingartner, G. Schurz (Eds.), Law and Prediction in the Light of Chaos Research. X, 291 pages. 1996.

Vol. 474: Aa. Sandqvist, P. O. Lindblad (Eds.), Barred Galaxies and Circumnuclear Activity. Proceedings of the Nobel Symposium 98, 1995. XI, 306 pages. 1996.

Vol. 475: J. Klamut, B. W. Veal, B. M. Dabrowski, P. W. Klamut, M. Kazimierski (Eds.), Recent Developments in High Temperature Superconductivity. Proceedings, 1995. XIII, 362 pages. 1996.

Vol. 476: J. Parisi, S. C. Müller, W. Zimmermann (Eds.), Nonlinear Physics of Complex Systems. Current Status and Future Trends. XIII, 388 pages. 1996.

Vol. 477: Z. Petru, J. Przystawa, K. Rapcewicz (Eds.), From Quantum Mechanics to Technology. Proceedings, 1996. IX, 379 pages. 1996.

Vol. 478: G. Sierra, M. A. Martín-Delgado (Eds.), Strongly Correlated Magnetic and Superconducting Systems. Proceedings, 1996. VIII, 323 pages. 1997.

Vol. 479: H. Latal, W. Schweiger (Eds.), Perturbative and Nonperturbative Aspects of Quantum Field Theory. Proceedings, 1996. X, 430 pages. 1997.

Vol. 480: H. Flyvbjerg, J. Hertz, M. H. Jensen, O. G. Mouritsen, K. Sneppen (Eds.), Physics of Biological Systems. From Molecules to Species. X, 364 pages. 1997.

Vol. 481: F. Lenz, H. Grießhammer, D. Stoll (Eds.), Lectures on QCD. VII, 276 pages. 1997.

Vol. 482: X.-W. Pan, D. H. Feng, M. Vallières (Eds.), Contemporary Nuclear Shell Models. Proceedings, 1996. XII, 309 pages. 1997.

Vol. 483: G. Trottet (Ed.), Coronal Physics from Radio and Space Observations. Proceedings, 1996. XVII, 226 pages. 1997.

Vol. 484: L. Schimansky-Geier, T. Pöschel (Eds.), Stochastic Dynamics. XVIII, 386 pages. 1997.

Vol. 485: H. Friedrich, B. Eckhardt (Eds.), Classical, Semiclassical and Quantum Dynamics in Atoms. VIII, 341 pages. 1997.

Vol. 486: G. Chavent, P. C. Sabatier (Eds.), Inverse Problems of Wave Propagation and Diffraction. Proceedings, 1996. XV, 379 pages. 1997.

Vol. 487: E. Meyer-Hofmeister, H. Spruik (Eds.), Accretion Disks – New Aspects. Proceedings, 1996. XIII, 356 pages. 1997.

Vol. 488: B. Apagyi, G. Endrédi, P. Lévay (Eds.), Inverse and Algebraic Quantum Scattering Theory. Proceedings, 1996. XV, 385 pages. 1997.

Vol. 489: G. M. Simnett, C. E. Alissandrakis, L. Vlahos (Eds.), Solar and Heliospheric Plasma Physics. Proceedings, 1996. VIII, 278 pages. 1997.

Vol. 490: P. Kutler, J. Flores, J.-J. Chattot (Eds.), Fifteenth International Conference on Numerical Methods in Fluid Dynamics. Proceedings, 1996. XIV, 653 pages. 1997.

Vol. 491: O. Boratav, A. Eden, A. Erzan (Eds.), Turbulence Modeling and Vortex Dynamics. Proceedings, 1996. XII, 245 pages. 1997.

Vol. 492: M. Rubí, C. Pérez-Vicente (Eds.), Complex Behaviour of Glassy Systems. Proceedings, 1996. IX, 467 pages. 1997.

Vol. 493: P. L. Garrido, J. Marro (Eds.), Fourth Granada Lectures in Computational Physics. XIV, 316 pages. 1997.

Vol. 494: J. W. Clark, M. L. Ristig (Eds.), Theory of Spin Lattices and Lattice Gauge Models. Proceedings, 1996. XI, 194 pages. 1997.

Vol. 495: Y. Kosmann-Schwarzbach, B. Grammaticos, K. M. Tamizhmani (Eds.), Integrability of Nonlinear Systems. XX, 404 pages. 1997.

Vol. 496: F. Lenz, H. Grießhammer, D. Stoll (Eds.), Lectures on QCD. VII, 483 pages. 1997.

Vol. 497: J.P. De Greve, R. Blomme, H. Hensberge (Eds.), Stellar Atmospheres: Theory and Observations. Proceedings, 1996. VIII, 352 pages. 1997.

Vol. 498: Z. Horváth, L. Palla (Eds.), Conformal Field Theories and Integrable Models. Proceedings, 1996. X, 251 pages. 1997.

Vol. 499: K. Jungmann, J. Kowalski, I. Reinhard, F. Träger (Eds.), Atomic Physics Methods in Modern Research. IX, 448 pages. 1997.

Vol. 500: D. Joubert (Ed.), Density Functionals: Theory and Applications. XVI, 194 pages. 1998.

Vol. 501: J. Kertész, I. Kondor (Eds.), Advances in Computer Simulation. VIII, 166 pages. 1998.

Vol. 502: H. Aratyn, T. D. Imbo, W.-Y. Keung, U. Sukhatme (Eds.), Supersymmetry and Integrable Models. XI, 379 pages. 1998.

New Series m: Monographs

Vol. m 2: P. Busch, P. J. Lahti, P. Mittelstaedt, The Quantum Theory of Measurement. XIII, 165 pages. 1991. Second Revised Edition: XIII, 181 pages. 1996.

Vol. m 3: A. Heck, J. M. Perdang (Eds.), Applying Fractals in Astronomy. IX, 210 pages. 1991.

Vol. m 4: R. K. Zeytounian, Mécanique des fluides fondamentale. XV, 615 pages, 1991.

Vol. m 5: R. K. Zeytounian, Meteorological Fluid Dynamics. XI, 346 pages. 1991.

Vol. m 6: N. M. J. Woodhouse, Special Relativity. VIII, 86 pages. 1992.

Vol. m 7: G. Morandi, The Role of Topology in Classical and Quantum Physics. XIII, 239 pages. 1992.

Vol. m 8: D. Funaro, Polynomial Approximation of Differential Equations. X, 305 pages. 1992.

Vol. m 9: M. Namiki, Stochastic Quantization. X, 217 pages. 1992.

Vol. m 10: J. Hoppe, Lectures on Integrable Systems. VII, 111 pages. 1992.

Vol. m 11: A. D. Yaghjian, Relativistic Dynamics of a Charged Sphere. XII, 115 pages. 1992.

Vol. m 12: G. Esposito, Quantum Gravity, Quantum Cosmology and Lorentzian Geometries. Second Corrected and Enlarged Edition. XVIII, 349 pages. 1994.

Vol. m 13: M. Klein, A. Knauf, Classical Planar Scattering by Coulombic Potentials. V, 142 pages. 1992.

Vol. m 14: A. Lerda, Anyons. XI, 138 pages. 1992.

Vol. m 15: N. Peters, B. Rogg (Eds.), Reduced Kinetic Mechanisms for Applications in Combustion Systems. X, 360 pages. 1993.

Vol. m 16: P. Christe, M. Henkel, Introduction to Conformal Invariance and Its Applications to Critical Phenomena. XV, 260 pages. 1993.

Vol. m 17: M. Schoen, Computer Simulation of Condensed Phases in Complex Geometries. X, 136 pages. 1993.

Vol. m 18: H. Carmichael, An Open Systems Approach to Quantum Optics. X, 179 pages. 1993.

Vol. m 19: S. D. Bogan, M. K. Hinders, Interface Effects in Elastic Wave Scattering. XII, 182 pages. 1994.

Vol. m 20: E. Abdalla, M. C. B. Abdalla, D. Dalmazi, A. Zadra, 2D-Gravity in Non-Critical Strings. IX, 319 pages. 1994.

Vol. m 21: G. P. Berman, E. N. Bulgakov, D. D. Holm, Crossover-Time in Quantum Boson and Spin Systems. XI, 268 pages. 1994.

Vol. m 22: M.-O. Hongler, Chaotic and Stochastic Behaviour in Automatic Production Lines. V, 85 pages. 1994.

Vol. m 23: V. S. Viswanath, G. Müller, The Recursion Method. X, 259 pages. 1994.

Vol. m 24: A. Ern, V. Giovangigli, Multicomponent Transport Algorithms. XIV, 427 pages. 1994.

Vol. m 25: A. V. Bogdanov, G. V. Dubrovskiy, M. P. Krutikov, D. V. Kulginov, V. M. Strelchenya, Interaction of Gases with Surfaces. XIV, 132 pages. 1995.

Vol. m 26: M. Dineykhan, G. V. Efimov, G. Ganbold, S. N. Nedelko, Oscillator Representation in Quantum Physics. IX, 279 pages. 1995.

Vol. m 27: J. T. Ottesen, Infinite Dimensional Groups and Algebras in Quantum Physics. IX, 218 pages. 1995.

Vol. m 28: O. Piguet, S. P. Sorella, Algebraic Renormalization. IX, 134 pages. 1995.

Vol. m 29: C. Bendjaballah, Introduction to Photon Communication. VII, 193 pages. 1995.

Vol. m 30: A. J. Greer, W. J. Kossler, Low Magnetic Fields in Anisotropic Superconductors. VII, 161 pages. 1995.

Vol. m 31: P. Busch, M. Grabowski, P. J. Lahti, Operational Quantum Physics. XI, 230 pages. 1995.

Vol. m 32: L. de Broglie, Diverses questions de mécanique et de thermodynamique classiques et relativistes. XII, 198 pages. 1995.

Vol. m 33: R. Alkofer, H. Reinhardt, Chiral Quark Dynamics. VIII, 115 pages. 1995.

Vol. m 34: R. Jost, Das Märchen vom Elfenbeinernen Turm. VIII, 286 pages. 1995.

Vol. m 35: E. Elizalde, Ten Physical Applications of Spectral Zeta Functions. XIV, 228 pages. 1995.

Vol. m 36: G. Dunne, Self-Dual Chern-Simons Theories. X, 217 pages. 1995.

Vol. m 37: S. Childress, A.D. Gilbert, Stretch, Twist, Fold: The Fast Dynamo. XI, 410 pages. 1995.

Vol. m 38: J. González, M. A. Martín-Delgado, G. Sierra, A. H. Vozmediano, Quantum Electron Liquids and High-T_c Superconductivity. X, 299 pages. 1995.

Vol. m 39: L. Pittner, Algebraic Foundations of Non-Commutative Differential Geometry and Quantum Groups. XII, 469 pages. 1996.

Vol. m 40: H.-J. Borchers, Translation Group and Particle Representations in Quantum Field Theory. VII, 131 pages. 1996.

Vol. m 41: B. K. Chakrabarti, A. Dutta, P. Sen, Quantum Ising Phases and Transitions in Transverse Ising Models. X, 204 pages. 1996.

Vol. m 42: P. Bouwknegt, J. McCarthy, K. Pilch, The W_3 Algebra. Modules, Semi-infinite Cohomology and BV Algebras. XI, 204 pages. 1996.

Vol. m 43: M. Schottenloher, A Mathematical Introduction to Conformal Field Theory. VIII, 142 pages. 1997.

Vol. m 44: A. Bach, Indistinguishable Classical Particles. VIII, 157 pages. 1997.

Vol. m 45: M. Ferrari, V. T. Granik, A. Imam, J. C. Nadeau (Eds.), Advances in Doublet Mechanics. XVI, 214 pages. 1997.

Vol. m 46: M. Camenzind, Les noyaux actifs de galaxies. XVIII, 218 pages. 1997.

Vol. m 47: L. M. Zubov, Nonlinear Theory of Dislocations and Disclinations in Elastic Body. VI, 205 pages. 1997.

Vol. m 48: P. Kopietz, Bosonization of Interacting Fermions in Arbitrary Dimensions. XII, 259 pages. 1997.

Vol. m 49: M. Zak, J. B. Zbilut, R. E. Meyers, From Instability to Intelligence. Complexity and Predictability in Nonlinear Dynamics. XIV, 552 pages. 1997.

Vol. m 50: J. Ambjørn, M. Carfora, A. Marzuoli, The Geometry of Dynamical Triangulations. VI, 197 pages. 1997.

Vol. m 51: G. Landi, An Introduction to Noncommutative Spaces and Their Geometries. XI, 200 pages. 1997.

Vol. m 52: M. Hénon, Generating Families in the Restricted Three-Body Problem. XI, 278 pages. 1997.

Vol. m53: M. Gad-el-Hak, A. Pollard, J.-P. Bonnet (Eds.), Flow Control. Fundamentals and Practices. XII, 527 pages. 1998.